Acta Hæmatologica

Founded 1948 by E. Meulengracht, K. Rohr, and G. Rosenow
Continued by H. Lüdin (1960–1977), U. Bucher (1977–1983),
R.S. Hillman (1978–1987), H.R. Marti (1983–1988), E.A. Beck (1984–1988)

KARGER

Printed in Switzerland
on acid-free paper by
Basler Zeitung AG, Basel

Appears 6-weekly:
2 volumes per year (8 issues)

Acta Hæmatologica

Submission: Only original papers written in English are considered and should be sent to:

Professor I. Ben-Bassat
S. Karger AG
Editorial Office 'Acta Haematologica'
P.O. Box
CH-4009 Basel (Switzerland)

Manuscripts should be submitted in triplicate (with three sets of illustrations of which one is an original), typewritten double-spaced on one side of the paper, with a wide margin.

Conditions: All manuscripts are subject to editorial review. Manuscripts are received with the explicit understanding that they are not under simultaneous consideration by any other publication. Submission of an article for publication implies the transfer of the copyright from the author to the publisher upon acceptance. Accepted papers become the permanent property of 'Acta Haematologica' and may not be reproduced by any means, in whole or in part, without the written consent of the publisher. It is the author's responsibility to obtain permission to reproduce illustrations, tables, etc. from other publications.

Arrangement:
Recommendation for Usage of Units and Abbreviations: The Editorial Board of 'Acta Haematologica' supports the recommendations of the 'International Committee for Standardization in Hematology' for using the SI (Système International d'Unités) equivalents in haematological literature. The standard abbreviations used for measurements of weight, length, volume, and time (i.e., mg, cm, ml, min) are used without definition. The abbreviations listed next page also require no preliminary definition. Please note that punctuation is not needed. When using other abbreviations, spell out the name in full at the first mention, and follow with the abbreviation in parentheses.
Title page: The first page of each paper should indicate the title (main title underlined), the authors' names, and the institute where the work was conducted. A short title for use as running head as well as the full address of the author to whom correspondence should be sent are also required.
Full address: The exact postal address complete with postal code must be given at the bottom of the title page. Please also supply Phone and Fax numbers.
Key words: For indexing purposes, a list of 3-10 key words in English is essential.
Abstract: Each paper needs an abstract of up to 10 lines.
Small type: Paragraphs which can or must be set in smaller type (case histories, test methods, etc.) should be indicated with a 'p' (petit) in the margin on the left-hand side.
Footnotes: Avoid footnotes. When essential, they are numbered consecutively and typed at the foot of the appropriate page.
Tables and illustrations: Tables and illustrations (both numbered in Arabic numerals) should be prepared on separate sheets. Tables require a heading and figures a legend, also prepared on a separate sheet. For the reproduction of illustrations, only good drawings and original photographs can be accepted; negatives or photocopies cannot be used. When possible, group several illustrations on one block for reproduction (max. size 180 × 223 mm) or provide crop marks. On the back of each illustration, indicate its number, the author's name, and 'top'. Color illustrations are reproduced at the author's expense.
References: In the text identify references by Arabic numerals [in square brackets]. Material submitted for publication but not yet accepted should be noted as 'unpublished data' and not be included in the reference list. The list of references should include only those publications which are cited in the text. Do not alphabetize; number references in the order in which they are first mentioned in the text. The surnames of the authors followed by initials should be given. There should be no punctuation other than a comma to separate the authors. Cite all authors, 'et al' is not sufficient. Abbreviate journal names according to the Index Medicus system. (Also see International Committee of Medical Journal Editors: Uniform requirements for manuscripts submitted to biomedical journals. Br Med J 1988;296:401-405.)
Examples
(a) Papers published in periodicals: Kauffman HF, van der Heide S, Beaumont F, Blok H, de Vries K: Class-specific antibody determination against *Aspergillus fumigatus* by means of the enzyme-linked immunosorbent assay. III. Comparative study: IgG, IgA, IgM ELISA titers, precipitating antibodies and IgE binding after fractionation of the antigen. Int Arch Allergy Appl Immunol 1986;80:300-306.
(b) Monographs: Matthews DE, Farewell VT: Using and Understanding Medical Statistics. Basel, Karger, 1985.
(c) Edited books: Hardy WD Jr, Essex M: FeLV-induced feline acquired immune deficiency syndrome: A model for human AIDS; in Klein E (ed): Acquired Immunodeficiency Syndrome. Prog Allergy. Basel, Karger, 1986, vol 37, pp 353-376.
Further details: Authors may consult our leaflet 'Rules for the Preparation of Manuscripts', available on request from the Publisher, S. Karger AG, P.O. Box, CH-4009 Basel (Switzerland).

Page charges: Authors are charged SFr. 115.- / US $ 75.00 per printed page for the first 6 pages and SFr. 225.- / US $ 150.00 for each additional page, including tables, illustrations, and references. 1 printed page is equal to approximately 3 manuscript pages.

Rapid communication: Manuscripts intended for rapid communication must present new findings of sufficient importance to justify their accelerated acceptance. They should follow the general arrangement of research papers except that the separate sections should not be formally titled. They should not be more than three pages in length (including figures, tables and references). Proofs are checked by the Main Editor and not sent to the authors. Review will be rapid, and once accepted, the paper will be included in the next planned issue. If manuscripts require significant change after editing they will be treated as brief communications and returned to the authors for corrections.

Brief communication: The general arrangement for brief communications is the same as for research papers except that the manuscript is not divided into separate sections formally entitled introduction, methods, results, discussion. No abstract is required. These should not be more than 3 pages in length.

Galley proofs: Unless indicated otherwise, galley proofs are sent to the first-named author and should be returned with the least possible delay. Alterations made in galley proofs, other than the correction of printer's errors, are charged to the author. No page proofs are supplied.

Reprints: Order forms and a price list are sent with the galley proofs. Orders submitted after the issue is printed are subject to considerably higher prices. Allow five weeks from date of publication for delivery of reprints.

KARGER

General Information

Acta Hæmatologica

Publication data: 'Acta Haematologica' is published 8 times annually. Volumes 89–90 with 4 issues each appear in 1993.
Subscription rates: Subscriptions run for a full calendar year. Prices are given per volume.
Personal subscription:
Subscription + postage and handling
SFr. 252.– + SFr. 11.20 Europe
SFr. 252.– + SFr. 16.80 Overseas
US$ 168.00 + US$ 11.20
(Must be in the name of, billed to, and paid by an individual. Order must be marked 'personal subscription'.)
Institutional subscription:
Subscription + postage and handling
SFr. 360.– + SFr. 16.– Europe
SFr. 360.– + SFr. 24.– Overseas
US$ 240.00 + US$ 16.00
Airmail postage: SFr. 24.–, US$ 20.00. (Extra per volume.)
Discount subscription prices:
International Society of Haematology
Microform edition: SFr. 180.–, US$ 120.00. (Available exclusively to subscribers of the paper edition.)
Advertising: Correspondence and rate requests should be addressed to the Publisher.

Subscription orders: Orders can be placed at agencies, bookstores, directly with the Publisher S. Karger AG, P.O. Box, CH – 4009 Basel (Switzerland) or with any of the following distributors:
Deutschland: S. Karger GmbH
Lörracher Strasse 16A, Postfach
D-W-7800 Freiburg
France: Librairie Luginbühl
36, bd de Latour-Maubourg, F-75007 Paris
Great Britain, Ireland: S. Karger Publishers
11 Ravenscroft Road
Beckenham, Kent BR3 4TP
USA: S. Karger Publishers, Inc.
26 West Avon Road, P.O. Box 529
Farmington, CT 06085
India, Bangladesh, Sri Lanka:
S. Karger India Subscription Agency
B – 5/132 Safdarjung Enclave
New Delhi-110029
Japan: Katakura Libri, Inc.
36-9 Hongo 3-chome, Bunkyo-ku, Tokyo 113
Change of address: Both old and new address should be sent to the subscription source.
Single issues and back volumes: Information can be obtained through the Publisher.

Bibliographic indices: This journal is regularly listed in bibliographic services, including *Current Contents*®.
Book Reviews: Books are accepted for review by special agreement.

KARGER

Usage of Units and Abbreviations

Entity	Abbreviation	Recommended SI equivalent	Not recommended
Red blood cell count	RBC	10^{12}/l	10^6/mm^3
White blood cell count	WBC	10^9/l	10^3/mm^3
Platelet count		... × 10^9/l	... × 10^3/mm^3
Reticulocyte count		% (or ... × 10^9/l)	‰ (or ../mm^3)
Haemoglobin	Hb	g/dl[a]	g/100 ml
Packed cell volume	PCV	ratio[b]	%
Mean cell volume	MCV	fl	
Mean cell haemoglobin	MCH	pg	
Mean cell haemoglobin concentration	MCHC	g/dl	%
Mean cell diameter	MCD	μm	
Sedimentation rate	ESR	mm in 1 h[c]	
Plasma haemoglobin	Plasma Hb	mg/l	mg/100 ml
Serum vitamin B_{12}		ng/l	
Serum folate		μg/l	
Serum iron		μmol/l	μg/100 ml
Total iron binding capacity	TIBC	μmol/l	μg/100 ml
Unsaturated iron binding capacity	UIBC	μmol/l	μg/100 ml
Transferrin		g/l	mg/100 ml
Serum haptoglobin		g/l	mg/100 ml
Plasma fibrinogen		g/l	mg/100 ml

[a] The concentration for Hb only is expressed as ../dl and not ../l or ../100 ml.
[b] No unit necessary, 1/1 is implied.
[c] Indicate method only if other than Westergren.

New Horizons with Biological Response Modifiers in Haematological Malignancies

New Horizons with Biological Response Modifiers in Haematological Malignancies

Satellite Symposium at the
24th Congress of the
International Society of
Haematology,
London, August 23–27, 1992

Guest Editor
J.M. Goldman, London

6 figures and 11 tables, 1993

KARGER

Acta Haematologica
Vol. 89, Supplement 1, 1993

S. Karger
Medical and Scientific Publishers
Basel·Freiburg·Paris·London
New York·New Delhi·Bangkok
Singapore·Tokyo·Sydney

Drug Dosage
The authors and the publisher have exerted every effort to ensure that drug selection and dosage set forth in this text are in accord with current recommendations and practice at the time of publication. However, in view of ongoing research, changes in government regulations, and the constant flow of information relating to drug therapy and drug reactions, the reader is urged to check the package insert for each drug for any change in indications and dosage and for added warnings and precautions. This is particularly important when the recommended agent is a new and/or infrequently employed drug.

Printed in Switzerland on acid-free paper by
Reinhardt Druck Basel
ISBN 3-8055-5801-5

Contents

Acta Haematol 1993;89(suppl 1):1

John M. Goldman

Editorial

Let me welcome you to this satellite symposium organized by F. Hoffmann-La Roche Ltd in conjunction with the meeting of the International Society of Haematology in London. The topic is biological response modifiers in haematological malignancies, which is most timely.

I suppose the term biological response modifier, which of course is rather unwieldy, is about 10 years old. I confess I was rather sceptical when it was introduced. It then seemed unlikely that any molecule of human or mammalian origin could be harnessed to modify biological responses in a clinically useful manner – remember that we were then still in the heyday of the cytotoxic era and many oncologists believed that cure of leukaemia and indeed of solid tumours depended simply on learning how to give bigger and better doses of existing chemotherapeutic agents. This attitude persists today – and provides a rationale for autografting – but few would deny that there are also subtler ways of controlling and probably of curing tumours.

I confess that my scepticism extended also to interferon-α (IFN-α) when it entered clinical practice. I was dubious as to whether it could really do all that was claimed for it in hairy cell leukaemia and dubious as to whether it could really be useful in chronic myelocytic leukaemia (CML). I frankly admit that I (and others) treated the first reports of Philadelphia chromosome negativity with incredulity. We 'knew' that drugs alone could rarely induce complete remission. Of course we were wrong. We know now that IFN-α induces remission in 10–20% of patients with previously untreated CML and some of these patients have no evidence of disease when blood and marrow are studied by the polymerase chain reaction – truly a remarkable finding. IFN-α responders survive longer than nonresponders. Today IFN-α plays a major role in the management of the CML patient. We will hear shortly about the value of IFN-α in CML relapsing after bone marrow transplantation and about IFN-α used in conjunction with cytosine arabinoside.

The importance of these results with IFN-α went beyond its benefit for patients with CML and hairy cell leukaemia, though this of course was important. They drew attention to the fact that malignant diseases can be treated by noncytotoxic methods. If IFN-α worked, then other biological molecules could work also. Hence the interest in interleukin-2 (IL-2). Originally identified as T-cell growth factor, we know now a great deal about the ligand-receptor interactions at cellular and molecular levels. Again it seemed intuitively a little improbable that IL-2 could be clinically useful, but again the critics were confounded. Today IL-2 seems to be useful in selected solid tumours and may have a real role to play in the management of acute leukaemia. We will hear later about its use to consolidate remission after autografting for acute myeloic leukaemia and acute lymphatic leukaemia.

Finally – differentiating factors. I have watched with some fascination the unfolding of the story in the lab. Dr. McCarthy in my group was one of the first to demonstrate that vitamin D_3 would induce differentiation in HL60 cells – a most dramatic and exciting observation – but clinically vitamin D_3 did little for patients with leukaemia or myelofibrosis. Other possible differentiating agents, cytosine arabinoside, DMSO or *cis*-retinoic acid, were also failures in the oncology clinic.

Many will remember the excitement that greeted the original report from Shanghai that all-*trans*-retinoic acid would induce complete remission in patients with acute promyelocytic leukaemia (AML-FAB M3). This was soon confirmed by the French group and we are lucky today to have a representative of the French APL Study Group to update with this intriguing story.

Prof. John M. Goldman
MRC/LRF Leukemia Unit, Hematology
Royal Postgraduate Medical School
Hammersmith Hospital
London W12 ONN (UK)

Acta Haematol 1993;89(suppl 1):2–7

A. Fefer, M. Benyunes
C. Higuchi, A. York
C. Massumoto, C. Lindgren
C.D. Buckner, J.A. Thompson

Division of Oncology,
University of Washington and the Fred Hutchinson Cancer Research Center,
Seattle, Wash., USA

Interleukin-2 ± Lymphocytes as Consolidative Immunotherapy after Autologous Bone Marrow Transplantation for Hematologic Malignancies

Key Words
Bone marrow transplantation
Immunotherapy
Interleukin-2
Leukemia
Lymphokine-activated killer cells
Lymphoma

Abstract
Patients who undergo autologous bone marrow transplantation (ABMT) for advanced hematologic malignancies experience a high relapse rate. Therapy with interleukin-2 (IL-2) ± lymphokine-activated killer (LAK) cells has induced clinical responses in some patients with advanced malignant lymphoma (ML) or acute myelogenous leukemia (AML). It is postulated that IL-2 ± LAK cells represents a potentially non-cross-resistant therapeutic modality which might prevent or delay relapses if used as consolidative immunotherapy after ABMT, at a time of minimal residual disease. Therefore, we first studied the reconstitution of IL-2-responsive LAK precursor cells after ABMT and found them in the circulation as early as 3 weeks after ABMT. A phase Ib clinical trial was then performed which identified a tolerable IL-2 regimen which could be administered early after ABMT and which could induce immunomodulatory effects. We then initiated a clinical trial to determine the feasibility of generating and administering autologous LAK cells using this IL-2 regimen after ABMT for 16 patients with ML. The results show that IL-2 + LAK therapy early after ABMT is feasible but is more toxic than IL-2 alone. Patients with AML on the phase I IL-2 trial and with ML on the IL-2 + LAK protocol were evaluated for tumor status. Of 8 patients with AML in first relapse or at a later stage who underwent ABMT and received IL-2, 2 have relapsed, while 6 remain in complete remission 26+ to 40+ (median 28+) months after ABMT. Of 16 patients with ML considered at high risk for relapse who were treated with ABMT + IL-2 + LAK, 5 have relapsed, while 11 remain in complete remission at 6+ to 21+ (median 10+) months after ABMT. The results in both trials are quite encouraging and appear to be better than those in nonrandomized historical controls at our institution. Prospectively randomized trials of IL-2 versus no IL-2 after ABMT in such patients are being initiated to assess definitively the effect, if any, on the relapse rate.

A. Fefer
Division of Oncology
University of Washington
Fred Hutchinson Cancer Research Center
Seattle, WA 98195 (USA)

Introduction

High-dose chemoradiotherapy and autologous bone marrow transplantation (ABMT) can cure *some* patients with acute leukemia or lymphoma refractory to conventional therapy [1]. However, ABMT for advanced leukemia or lymphoma, i.e. beyond the first complete remission (CR), is associated with a low probability of disease-free survival, largely due to a 40–80% probability of relapse. For example, in a series of 101 patients with malignant lymphoma (ML) who underwent ABMT in Seattle, the group exhibited an 84% probability of relapse, with a median time to relapse of only 2–4 months after bone marrow transplantation (BMT) [2]. Attempts to use additional chemotherapy or radiation to decrease the relapse rate have been hampered by the cross-resistance of the tumor to the various agents and by their shared cumulative side effects.

If therapy with interleukin-2 (IL-2) ± lymphokine-activated killer (LAK) cells were to be used after BMT, it would have to be administered late enough for the patient to have established a graft and to have recovered from the major BMT-related toxicities, but early enough before the malignancy would be likely to recur. Moreover, IL-2-responsive LAK precursor cells would have to be present in the circulation at that time.

This article will briefly summarize the results of a phase Ib clinical trial of IL-2 after ABMT for hematologic malignancies, present preliminary data on the generation and infusion of autologous LAK cells in such a setting, and present clinical follow-up results of IL-2-based immunotherapy after ABMT for hematologic malignancy – results which, although preliminary, are sufficiently encouraging to warrant prospective randomized trials.

Table 1. Design of phase Ib trial of IL-2 after ABMT

Group	Patients n	'Induction' dose $\times 10^6$ U/m^2/day (days 1–5)	'Maintenance' dose $\times 10^6$ U/m^2/day (days 12–21)
I	4	0.3	0.3
II	4	1.0	0.3
III	6	3.0	0.3
IV	2	4.5	0.3

Table 2. Design of IL-2+ LAK protocol

	Days
IL-2 (3×10^6 U/m^2/day)	1–5
Pheresis	7–9
LAK infusion	12–14
IL-2 (3×10^5 U/m^2/day)	12–21

Table 3. ABMT + IL-2 for AML in ≥first relapse vs. historical controls (ABMT without IL-2)

Group	Relapsed		Alive in CR	
	n	months	n	months
IL-2 (n=8)	2	4, 10	6	26+ to 40+ (median 28)
No IL-2 (n=15)[a]	9	2–16 (median 4)	4	7+ to 36+ (median 26+)

CR = Complete remission.
[a] Two died of infections at 6 and 11 months.

Materials and Methods

Patient Selection

For the phase Ib clinical trial of IL-2 after ABMT, 16 patients who had undergone ABMT for a variety of hematologic malignancies including 8 with acute myelogenous leukemia (AML) in first relapse or beyond began IL-2 therapy a median of 33 days (range 14–91) after ABMT [3]. Patients had had a variety of pretransplant conditioning and marrow purging regimens. All had recovered from ABMT-associated acute toxicities and had trilineage marrow engraftment documented by marrow aspiration. Eligibility requirements included: adequate bone marrow function (a neutrophil count ≥500/ml; platelets >20,000/ml with or without transfusion support); adequate renal and hepatic function and absence of infection off antibiotics. For the IL-2+LAK trial, 16 patients with ML were treated. They were heterogeneous in regards to diagnosis (12 with non-Hodgkin's lymphoma and 4 with Hodgkin's disease), stage at ABMT (7 less advanced, i.e. in first relapse, and 9 advanced, i.e. second relapse or later), conditioning regimen (4 with chemotherapy alone and 12 with chemotherapy and total body irradiation). The marrow from patients with non-Hodgkin's lymphoma were all purged with a cocktail of anti-B cell antibodies prior to reinfusion. All protocols and consent forms were approved by the Institutional Review Committee of the Fred Hutchinson Cancer Research Center.

Design of Phase Ib Trial of IL-2 (table 1)

Patients were sequentially assigned to escalating IL-2 'induction' doses of $0.3–4.5 \times 10^6$ U/m^2/day on days 1–5 of the IL-2 protocol. After a rest period, all patients received a low-dose 'maintenance' IL-2 at 0.3×10^6 U/m^2/day on days 12–21. Induction IL-2 was administered in the hospital and maintenance IL-2 in the outpatient clinic.

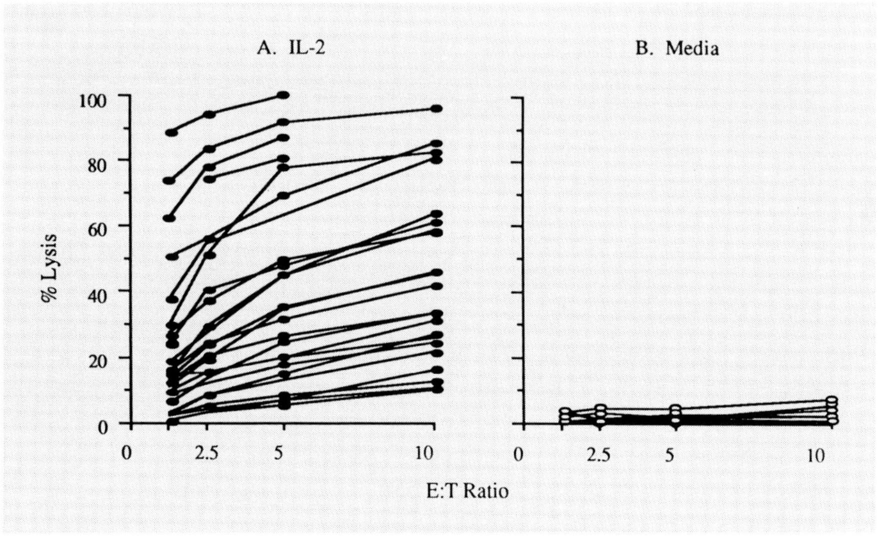

Fig. 1. LAK precursor activity is detectable after ABMT. Peripheral blood mononuclear cells obtained from 21 patients with refractory hematologic malignancies who were at various intervals within 3–12 weeks after ABMT were cultured for 4 or 5 days with IL-2 (1,000 U/ml, **A**), or culture media without IL-2 supplementation (**B**). Cells were assessed for cytolytic reactivity against Daudi.

Design of Feasibility Trial of IL-2+Autologous LAK Cells (table 2)

IL-2 at the maximal tolerated dose of 3×10^6 U/m^2/day was administered by continuous intravenous infusion on days 1–5 of the IL-2 protocol. The patients underwent leukapheresis daily on days 7–9. The cells were incubated with IL-2, 1,000 U/ml for 5 days, and the resultant LAK cells were infused daily on days 12–14. A low maintenance dose of IL-2 at 3×10^5 U/m^2/day was administered on days 12–21. The details of pheresis, LAK cell generation and LAK cell harvesting have been previously described [4].

Recombinant IL-2

Recombinant human IL-2 (provided by Hoffmann-La Roche, Inc., Nutley, N.J., USA) has a specific activity of $1.2–1.5 \times 10^7$ BRMP U/mg of protein. IL-2 was supplied as a lyophilized powder containing 100 mg of IL-2, with 25 mg of human serum albumin/10^6 units of IL-2, and 5 mg of mannitol/10^6 units of IL-2. The powder was reconstituted in sterile saline for injection.

Immunologic Monitoring

Immunologic monitoring, performed on all patients, consisted of the flow cytometric analysis of a patient's lymphocyte phenotype using a panel of monoclonal antibodies, and the assessment of natural killer and LAK cytotoxicity. Patients were studied before IL-2 therapy, 1 day after the 5-day induction course of IL-2, and 1 day after completing the maintenance course of IL-2. All assays and methods have been previously described [5]. LAK activity was defined by the ability of the patient's freshly isolated peripheral blood mononuclear cells to lyse natural killer-resistant Daudi targets in a 4-hour ^{51}Cr-release assay.

Statistical Analysis

Product limit estimates of the probability of relapse and disease-free survival for IL-2 ± LAK-treated patients were derived according to the Kaplan-Meier method [6].

Results

Reconstitution of LAK Precursor Cells after ABMT

Peripheral blood mononuclear cells obtained from 21 patients 17–83 days after ABMT were cultured with IL-2 at 1,000–2,000 U/ml for 4–5 days and tested for their ability to lyse Daudi at different effector:target ratios [7]. The results are shown in figure 1. Each curve represents the lysis by cells from a single patient obtained at one time and incubated with IL-2 or in control media. LAK precursor activity was detectable in the circulation of virtually every patient tested at every time. Positive and negative selection experiments on the FACS revealed that the LAK activity was mediated largely by $CD3^-$ $CD56^+$ cells with a minor subset of cells which stained dimly for CD8 and coexpressed CD56 [7, 8].

Phase Ib Trial of IL-2 after ABMT for Hematologic Malignancies

The design of the phase Ib trial (table 1) [3] was based on experience with IL-2 for treating patients with renal cell carcinoma [9], with a view to adding LAK cells to the regimen once the maximal tolerated dose of IL-2 was identified. The common toxicities included fever, rash, nausea, diarrhea, elevation of bilirubin and creatinine, weight gain and, occasionally, dyspnea. All toxicities were reversible by stopping IL-2. Reversible hypotension and thrombocytopenia were the dose-limiting side effects. The maximum tolerated 'induction' dose of IL-2 was found to be 3.0×10^6 U/m^2/day. The IL-2 regimen induced an initial lymphopenia followed by a marked 'rebound' lymphocytosis representing a significant increase in cells expressing CD16 and CD56, with a concomitant increase in circulating LAK activity [3].

Feasibility Trial of IL-2+Autologous LAK Cells after ABMT for ML

Patients on this trial exhibited the same toxicity during the induction IL-2 cycle as did patients receiving the same IL-2 dose on the reported phase I IL-2 trial. Maintenance IL-2 therapy was well tolerated. Leukapheresis was associated largely with thrombocytopenia. A median of 136×10^9 cells were infused per patient. The cells infused were predominantly $CD3^+$, $CD8^+$ and $CD56^+$. LAK cell infusions were generally well tolerated, with only transient fever, rigors, and dyspnea which rapidly reversed. Hematologically, IL-2+LAK therapy was associated with modest neutrophilia, significant thrombocytopenia and marked lymphocytosis, all quite comparable to those observed with IL-2 therapy alone without LAK cells, except for thrombocytopenia, which was clearly more significant during the pheresis days. The immunologic changes observed were also comparable to those previously reported with IL-2 alone [10].

Clinical Follow-Up of Patients Treated with ABMT+IL-2 for AML

The clinical characteristics of the 8 patients have been described [3]. The duration of their first complete remission was 11 months (median). Five patients were conditioned with busulfan 8 mg/kg, cyclophosphamide 60 mg/kg, and TBI 21.0 Gy, and 3 with busulfan 16 mg/kg and cyclophosphamide 200 mg/kg. Marrow from 4 patients was treated with 4-hydroxycyclophosphamide (4-HC) prior to cryopreservation. Four of the 8 patients received the IL-2 at the maximal tolerated dose, while 4 received lower doses of IL-2. Two patients have relapsed at 4 and 10 months, while 6 remain in complete remission 26+ to 40+ months after ABMT.

Clinical Characteristics and Follow-Up of the Historical Control Group

In the absence of a randomized control group, a historical control group was generated which consisted of all patients with AML in first relapse or beyond who underwent ABMT at our institution during the same time period with the same preparative and purging regimens as the IL-2 group, who met the eligibility criteria for IL-2 therapy but did not receive IL-2 either because they refused, or because IL-2 was temporarily unavailable to them. No patient who was eligible for IL-2 therapy was excluded from the control group. The 15 controls thus selected did not differ significantly from the 8 patients who received IL-2 with respect to age at ABMT, FAB class, number of cycles of induction and consolidation chemotherapy, duration of first complete remission, use of 4-HC-treated marrow, type of preparative regimen or time to engraftment. Table 3 compares the results of the 'controls' with those of the 'experimental' group. Two of the 15 control patients died of infections at 6 and 11 months, 9 relapsed, with a median of 4 months, while 4 remain in complete remission 7+ to 36+ (median 26+) months after ABMT.

Clinical Follow-Up of 16 Patients Treated with ABMT and IL-2+LAK for ML

Of the 16 patients treated in first relapse or beyond beginning a median of 31 days after ABMT, 5 relapsed (3 at 5 months, 1 at 8 months and 1 at 19 months after ABMT), while 11 remain in complete remission at 6+ to 21+ months (median 10+). These preliminary results compare favorably with our past experience in similar patients [2].

Discussion

Patients undergoing ABMT for advanced hematologic malignancy (e.g. with AML and ML in relapse) relapse often and early. The possibility that IL-2±lymphocytes after ABMT may decrease the relapse rate is being considered based on the following rationale: (a) human leukemia and lymphoma cells can be lysed in vitro by LAK cells [11, 12]; (b) the therapy may be non-cross-resistant with chemotherapy [13]; (c) in phase I/II trials, treatment with IL-2±LAK cells has induced partial or, rarely, complete remissions in *some* patients with hematologic malignancies [14–16]; (d) immunotherapy should theoretically be more effective against minimal residual disease, a setting readily induced by the conditioning regimens used for ABMT, and (e) therapy with IL-2±LAK cells after ABMT should theoretically be able to eradicate whatever clonogenic malignant cells might be contaminating the reinfused marrow and thus should obviate the need for purging the marrow of tumor cells [17]. Although strong circumstantial evidence for a GVL effect appears to exist only in recipients of *allogeneic* marrow [17] – especially those who exhibit graft-versus-host disease – one cannot rule out the possibility that a GVL effect also exists in recipients of *autologous* marrow and can be amplified by IL-2 and/or cells.

To be effective, such therapy would have to be administered early after ABMT, after the patient has recovered from ABMT toxicity, after IL-2-responsive LAK precursor cells have been reconstituted, and before relapses are likely to occur. LAK precursor cells in the blood have been detected as early as 14 days after ABMT [7, 18]. A phase Ib trial has identified an IL-2 regimen which is tolerable and induces immunomodulatory effects early after ABMT [3]. Long-term clinical follow-up of 8 patients with AML in first relapse or beyond on that IL-2 trial yielded very encouraging results, with 6 patients remaining in complete remission 26+ to 40+ months. These results are better than would be expected from previous reports [19–22] and compare favorably with results reported after allogeneic BMT for AML beyond first complete remission [23]. Moreover, they are significantly better than those in our concurrently transplanted historical controls who met the eligibility criteria for IL-2 but did not receive it because of patient refusal or IL-2 nonavailability.

Since the combination of IL-2+LAK cells has been therapeutically more effective than IL-2 alone in animal models [24, 25], it is possible that exogenous LAK cells might contribute to the antitumor activity of IL-2 after ABMT. Accordingly, a phase I/II feasibility trial of IL-2+LAK cells was performed in patients who had undergone ABMT for ML to determine whether LAK cells could be generated and infused in this setting [10]. The results clearly demonstrated that peripheral lymphocytosis was induced by the induction cycle of IL-2 and that a large number of LAK cells could be generated in vitro and infused in vivo. The most significant toxicity was thrombocytopenia, especially on the pheresis days. The additional toxicity associated with the pheresis and LAK cell infusions was sufficiently great to lead to the conclusion that although it is feasible, it is not recommended.

The ML patients in this trial were considered to be at a high risk of post-transplant relapse either because they were beyond second complete remission or because their disease was not responding to chemotherapy at the time of ABMT. Nevertheless, 11 of 16 patients remain in complete remission 6+ to 21+ (median 10+) months after ABMT. Despite the heterogeneity of the patients, these preliminary clinical results are encouraging, given that, historically, such patients have had a 60–80% probability of early relapse (median 3–4 months) [2].

Overall, these preliminary clinical results of IL-2 for AML and IL-2+LAK for ML after ABMT suggest that consolidative immunotherapy with IL-2 (±LAK cells) administered early after ABMT for patients with AML or ML known to be at high risk for early relapse may decrease the relapse rate. The data provide a basis for initiating prospectively randomized controlled trials of IL-2 versus no IL-2 after ABMT to establish definitively whether this approach will, indeed, reduce the relapse rate and ultimately improve disease-free survival in such patients.

Acknowledgements

Supported in part by Grants PO1 CA-47748, CA-18029, CA-15704 and T32 CA-09515, awarded by the National Cancer Institute, Department of Health and Human Services, and the Jose Carreras International Leukemia Foundation.

The authors wish to thank the following for their assistance in performing these studies: K. Beach, A. York, and B. Bolonesi for their work in coordinating the clinical care of the patients participating in these studies; G. Schoch for his expert assistance in the statistical analysis of the data presented; J. Factor, A.E. Rogers and H. Todd for their expert technical assistance, and Hoffmann-La Roche, Inc. for generously providing the IL-2.

References

1 Fefer A, Sullivan K: Bone marrow transplantation for hematologic malignancies; in Calabresi, Schein (eds): Medical Oncology: Basic Principles and Clinical Management of Cancer. Elmsford, Pergamon Press, in press.

2 Petersen F, Appelbaum F, Hill R, et al: Autologous marrow transplantation for malignant lymphoma: A report of 101 cases from Seattle. J Clin Oncol 1990;8:638.

3 Higuchi CM, Thompson JA, Petersen FB, et al: Toxicity and immunomodulatory effects of interleukin 2 after autologous bone marrow transplantation for hematologic malignancies. Blood 1991;77:2561.

4 Thompson J, Shulman K, Benyunes M, et al: Prolonged continuous intravenous infusion interleukin-2 and lymphokine-activated killer cell therapy for metastatic renal cell carcinoma. J Clin Oncol 1992;10:960.

5 Higuchi C, Triesman J, Thompson J, et al: Cytolytic T-lymphocytes infiltrating human melanoma expanded by culture with interleukin-2 (abstract 1592). Proc Am Assoc Cancer Res 1990;31:269.

6 Kaplan E, Meier P: Non-parametric estimation from incomplete observations. J Am Stat Assoc 1958;53:457.

7 Higuchi C, Thompson J, Cox T, et al: Lymphokine-activated killer function following autologous bone marrow transplantation for refractory hematologic malignancies. Cancer Res 1989;49:5509.

8 Ortaldo JR, Mason A, Overton R: Lymphokine-activated killer cells. J Exp Med 1986; 164:1193.

9 Thompson JA, Lee DJ, Lindgren CG, et al: Influence of schedule of interleukin 2 administration on therapy with interleukin 2 and lymphokine activated killer cells. Cancer Res 1989; 49:235.

10 Benyunes M, York A, Lindgren C, et al: IL-2±LAK cells as consolidative therapy after autologous BMT for hematologic malignancies: A feasibility trial. Proc Am Soc Clin Oncol 1992;11:319.

11 Oshimi K, Oshimi Y, Akutsu M, et al: Cytotoxicity of interleukin 2-activated lymphocytes for leukemia and lymphoma cells. Blood 1986; 68:938.

12 Adler A, Chervenick P, Whiteside T, et al: Interleukin-2 induction of lymphokine-activated killer (LAK) activity in the peripheral blood and bone marrow of acute leukemia patients. I. Feasibility of LAK generation in adult patients with active disease and in remission. Blood 1988;71:709.

13 Allavena P, Damia G, Colombo T, et al: Lymphokine-activated killer (LAK) and monocyte-mediated cytotoxicity on tumor cell lines resistant to antitumor agents. Cell Immunol 1989; 120:250.

14 Foa R, Meloni G, Tosti S, et al: Treatment of residual disease in acute leukemia patients with recombinant interleukin 2 (IL2): Clinical and biological findings. Bone Marrow Transplant 1990;6:98.

15 Rosenberg SA, Lotze MT, Muul LM, et al: A progress report on the treatment of 157 patients with advanced cancer using lymphokine-activated killer cells and interleukin-2 or high-dose interleukin-2 alone. N Engl J Med 1987;316:889.

16 West WH, Tauer KW, Yannelli JR, et al: Constant-infusion recombinant interleukin-2 in adoptive immunotherapy of advanced cancer. N Engl J Med 1987;316:898.

17 Fefer A, Truitt RL, Sullivan KM: Adoptive cellular therapy: Graft-vs-tumor responses after bone marrow transplantation; in DeVita, Hellmann, Rosenbert (eds): Biologic Therapy of Cancer: Principles and Practice. Philadelphia, Lippincott, 1991.

18 Keever CA, Welte K, Small T, et al: Interleukin 2-activated killer cells in patients following transplants of soybean lectin-separated and E rosette-depleted bone marrow. Blood 1987;70: 1893.

19 Stewart P, Buckner C, Bensinger W, et al: Autologous marrow transplantation in patients with acute nonlymphocytic leukemia in first remission. Exp Hematol 1985;13:267.

20 Yeager A, Kaizer H, Santos G, et al: Autologous bone marrow transplantation in patients with acute nonlymphocytic leukemia, using ex vivo marrow treatment with 4-hydroperoxycyclophosphamide. N Engl J Med 1986;315:142.

21 Chopra R, Goldstone A, McMillan A, et al: Successful treatment of acute myeloid leukemia beyond first remission with autologous bone marrow transplantation using busulfan/cyclophosphamide and unpurged marrow: The British Autograft Group experience. J Clin Oncol 1991;9:1840.

22 Gorin N, Aegerter P, Auvert B, et al: Autologous bone marrow transplantation for acute myelocytic leukemia in first remission: A European survey of the role of marrow purging. Blood 1990;75:1606.

23 Clift R, Buckner C, Thomas E, et al: The treatment of acute non-lymphoblastic leukemia by allogeneic marrow transplantation. Bone Marrow Transplant 1987;2:243.

24 Charak BS, Brynes RK, Groshen S, et al: Bone marrow transplantation with interleukin-2-activated bone marrow followed by interleukin-2 therapy for acute myeloid leukemia in mice. Blood 1990;76:2187.

25 Mulé J, Shu S, Rosenberg S: The anti-tumor efficacy of lymphokine-activated killer cells and recombinant interleukin 2 in vivo. J Immunol 1985;135:646.

Acta Haematol 1993;89(suppl 1):8–14

C.S. Higano
W.H. Raskind
J.W. Singer

University of Washington,
Fred Hutchinson Cancer Research Center,
Cell Therapeutics Inc.,
Seattle, Wash., USA

Use of Interferon Alfa-2a to Treat Hematologic Relapse of Chronic Myelogenous Leukemia after Bone Marrow Transplantation

Key Words
Allogeneic bone marrow transplantation
Alpha interferon
Chronic myelogenous leukemia
Immunotherapy
Interferon alfa-2a

Abstract
The use of recombinant human alpha interferon (IFN) to treat relapse of chronic myelogenous leukemia (CML) in chronic phase after bone marrow transplantation (BMT) was studied in a prospective trial. Relapse was defined as >90% metaphases containing the Philadelphia chromosome (Ph) and hematologic abnormalities consistent with chronic phase CML. In the initial 18 patients, all received allogeneic marrow, 17 from a related donor, 1 from an unrelated donor. Only 1 patient received T-depleted marrow initially. Three patients had relapsed after second BMT. IFN was started at 3×10^6 IU/m^2/day and escalated to the maximum tolerated dose or to a maximum of 6×10^6 IU/m^2/day. Elevated white blood counts and platelet counts were controlled in 14 of 16 and 6 of 6 patients, respectively. Six patients (33%) have had a complete disappearance of the Ph (cytogenetic complete response) and 2 have had a partial response (cytogenetic partial response <35% Ph$^+$ metaphases but >0%) on at least one sample. Six patients had no significant response after 9–12 months and 4 patients developed clinical accelerated phase or blast crisis after 3–6 months. IFN controlled the blood counts in 75% of patients. Of 4 patients with a sex marker, the Ph$^-$ population was of donor origin in 3 and of host origin in 1. Clonal cytogenetic abnormalities other than the Ph were present in 13 patients and did not predict for lack of response to IFN. IFN effectively produces hematologic control in the majority of patients and suppresses the Ph clone in up to one third. A trial of IFN treatment in an earlier stage of relapse is indicated.

Introduction

Bone marrow transplantation (BMT) is the only curative treatment for patients with chronic myelogenous leukemia (CML). Clinical relapse of CML after BMT remains a problem in approximately 7–30% of patients transplanted in chronic phase (CP) [1, 2]. Although second BMT is an option for selected patients who relapse late after initial BMT, it is associated with a high transplant-related mortality and relapse rate [3]. Most patients who relapse with CP CML after BMT receive therapy to control peripheral blood counts.

In newly diagnosed CP CML, alpha interferon (IFN) controls peripheral counts in 75% of patients and com-

Celestia S. Higano
University of Washington (RC-08)
1959 NE Pacific Street
Seattle, WA 48195 (USA)

0001–5792/93/0895–0008
$ 2.75/0

pletely suppresses the malignant clone, detected by the presence of the Philadelphia chromosome (Ph) in myeloid cells, in up to 20% of patients [4]. Although there is a suggestion that survival is improved with IFN therapy, this remains unproved. Because of the ability of IFN to control the hematologic abnormalities in the majority of patients, to suppress the Ph clone and perhaps impact survival, a phase I–II trial of IFN was undertaken in patients who relapsed with CML in CP after allogeneic BMT.

Methods

Patients

Relapse of CML in CP after BMT was required for study entry. Relapse was defined as >90% Ph^+ metaphases in bone marrow with hematologic abnormalities consistent with CP CML. Patients with greater than 20% blasts and promyelocytes in the marrow, clinical evidence of accelerated phase such as bone pain, fever, or a rising percentage of blasts in the peripheral smear were excluded from the study. Patients with additional clonal cytogenetic abnormalities (including +8, 17q–, or an additional Ph) who were otherwise in clinical CP were not excluded. Patients who required myelosuppressive therapy for control of chronic graft-versus-host disease (GvHD) were excluded. All patients were required to have serum bilirubin ≤2.0 mg/dl, serum creatinine ≤1.6 mg/dl, normal cardiac and neurologic function. Signed informed consent was required before patients were treated on protocol approved by the Investigational Review Board of the Fred Hutchinson Cancer Center.

Treatment

Interferon. Hoffmann-La Roche (Nutley, N.J., USA) supplied interferon alfa-2a (Roferon®-A). Patients were taught how to reconstitute the lyophilized drug and to self-administer subcutaneous injections of IFN.

Induction. IFN was started at a dose of 3.0×10^6 IU/m²/day. After 1 month, the dose was to be escalated to 4.5×10^6 IU/m²/day and to 6.0×10^6 IU/m²/day after 2 months as tolerated. The maximally tolerated dose was to be continued for at least 1 year.

Dose Modifications. The IFN dose was reduced by 50% if the performance status less than 80% on the Karnofsky scale, the SGOT or SGPT increased more than 5 times the control value, the bilirubin was greater than 2.5 mg/dl, the serum creatinine was greater than 2.2 mg/dl, the platelet count was less than 50×10^9/l, the abolute neutrophil count was less than 0.8×10^9/l, or weight loss was greater than 5% ideal body weight. If significant neurological toxicity developed, IFN was stopped for 2 weeks and restarted at 50% of the previous dose.

Off Study Criteria. If hematologic control could not be maintained with IFN alone or if grade 3 or 4 toxicity of any nature was persistent, the patient was taken off study. If, after 1 year of IFN therapy, there was no significant cytogenetic response (i.e., greater than 90% Ph^+ metaphases), patients were taken off study.

Maintenance. If complete disappearance of Ph^+ cells was achieved and maintained for at least 6 months, patients were to receive a maintenance dose of 1.5×10^6 IU/m² 3 times per week for at least 1 year. The dose and schedule were adjusted, when necessary, according to cytogenetic information.

Follow-Up Studies. The complete blood count was monitored weekly during dose escalation. After the maximum tolerated dose was established, the complete blood count was checked every 2–4 weeks thereafter at the discretion of the primary physician. Bone marrow aspirations and peripheral blood were requested every 3–4 months and sent to Seattle via overnight mail. Southern blot analysis for the BCR/ABL rearrangement [5] (Lifecodes Corp., Valhalla, N.Y., USA) was performed in some patients.

Y Body Analysis. The Y body was assayed in selected patients by in situ hybridization [6].

Reverse Transcription/Polymerase Chain Reaction Amplification. In a limited number of patients, reverse transcription/polymerase chain reaction for BCR/ABL was performed according to a modification [7] of the protocol described by Kawasaki et al. [8].

Response Criteria. In order to be in complete hematologic remission the white blood cell count (WBC) had to be less than 9×10^9/l with no immature forms (blasts, promyelocytes, myelocytes, metamyelocytes) and platelet count less than 450×10^9/l. Complete cytogenetic response is defined as complete disappearance of the Ph on at least one test. Partial response is defined as reduction of the Ph^+ population to less than 35% of metaphases but greater than 0%.

Results

Patients Characteristics

Nineteen patients were enrolled from May 1987 to January 1991. One patient was not eligible because chronic GvHD could not be controlled without azathioprine. Tables 1 and 2 summarize patient and transplant characteristics. Three patients had received IFN prior to BMT. In 2 patients IFN controlled the blood counts but there was no cytogenetic response in the 1 patient tested. The 3rd patient was on IFN only briefly before BMT.

Response

Hematologic Response. Sixteen patients had an elevated WBC ranging from 11.5 to 98.2×10^9/l with a median of 18.3×10^9/l and 6 patients had an elevated platelet count ranging from 547 to $1{,}500 \times 10^9$/l. The WBC was controlled in 14 patients including 6 patients with no cytogenetic response. IFN controlled the platelet count in all 6 patients despite the fact that 5 had no cytogenetic response. IFN did not control the WBC in 4 patients who went on to develop accelerated phase/blast crisis during the first 3–6 months on therapy.

The median pretreatment WBC and platelet counts of patients who achieved a cytogenetic response were 15.0×10^9/l and 240×10^9/l, respectively, whereas those of nonresponders were 27.5×10^9/l and 368×10^9/l, respectively. Only 1 of the 6 patients who presented with an elevated platelet count achieved a complete cytogenetic response.

Table 1. Patient characteristics (n = 18)

Male:female	11:7
Median age	38 (3–55)
Months[1] diagnosis to BMT	11 (1–82)
Relapse in chronic phase	18
Months[1] to 1st relapse	11 (1–48)
Months[1] to study entry	3 (0–41)
Patients with 2nd BMT	3

[1] Median values.

Table 2. Transplant characteristics (n = 18)

Stage at BMT	
Chronic	12
Accelerated	3
Blast crisis	3
Allogeneic donor	
Related, matched	16
Related, 1 antigen mismatch	1
Unrelated	1
Conditioning regimen	
Cy/TBI	15
Bu/Cy	1
Bu/Cy/TBI	1
VP-16/TBI	1
GvHD prophylaxis	
Csp/MTX	14
Csp/Pred	1
Csp	1
MTX	1
T depletion	1

Bu = Busulfan; Cy = cyclophosphamide; MTX = methotrexate; Pred = prednisone; TBI = total body irradiation; VP-16 = etoposide; Csp = cyclosporin A.

The median time to control WBC in all patients was 21 days (range 3–111 days). The rate of WBC response in patients with a cytogenetic response was similar (median 21 days, range 3–89) to that of nonresponders (median 25 days, range 7–111). Elevated platelet counts normalized in a median of 24 days of therapy (range 14–48 days). One patient developed an abrupt marrow aplasia 24 days after IFN was started. Although blast crisis was suspected, a complete cytogenetic response was documented. Marrow recovery occurred over 2 weeks' time.

Cytogenetic Response. Of 18 evaluable patients, 6 have had complete cytogenetic response on at least one sample, and 2 patients have had a partial response. Figure 1 demonstrates the time course of cytogenetic responses for all 8 responding patients. Ten patients were nonresponders: 6 did not respond after 9–12 months of IFN and 4 patients developed clinical accelerated phase/blast crisis after 3–6 months of therapy.

Two patients with complete cytogenetic responses have died. One patient developed blast crisis despite having only 35% Ph^+ metaphases and died during a second BMT. IFN was stopped intermittently in the other patient because of complex medical problems and she eventually died in blast crisis.

The median time to achieve a complete cytogenetic response was 3 months (range 1–18) with a median duration of response of 22+ months (range 6–31+). Four patients who were complete cytogenetic responders had no evidence of the BCR/ABL rearrangement by Southern blot analysis of peripheral blood or marrow. Testing was not performed in 2 patients.

Two patients achieved a partial cytogenetic response. One patient had relapsed after the second BMT and maintained partial response for 24 months. As the Ph^+ population increased and the patient became symptomatic of accelerated CML, she was treated with IFN plus donor buffy coat. Although the patient had severe three-system GvHD and still requires therapy 1 year after buffy coat infusion, the marrow remains Ph^-. After 6 months on IFN, the other patient had a decline from 100 to 27% Ph^+ metaphase cells and has since cycled between 27 and 66% Ph^+ metaphases.

In 4 responding patients with opposite sex donors, the Ph^- population was of donor origin in 3 patients and host in 1. In the latter patient, normal host and Ph^+ host cells without detectable donor (female) cells were present at the time IFN was started. After treatment with IFN the Ph^+ clone disappeared and only male, Ph^- host cells were detected. In situ hybridization with a Y-specific probe confirmed that at least 96% of peripheral blood mononuclear cells and at least 94% of peripheral blood granulocytes were host (male)-derived.

At study entry, 9 patients had clonal structural chromosomal aberrations detected in addition to the t(9;22). Cells of 4 additional patients contained clonal structural abnormalities at some time during IFN treatment. It is likely that these structural changes, likely due to the conditioning regimen, were present before IFN therapy was started but were not previously detected. Two patients had trisomy 8. Despite the presence of at least one clonal chromosomal abnormality in addition to Ph, 4 of these patients, including 1 with trisomy 8, achieved a complete cytogenetic remission and 1 patient had a partial response.

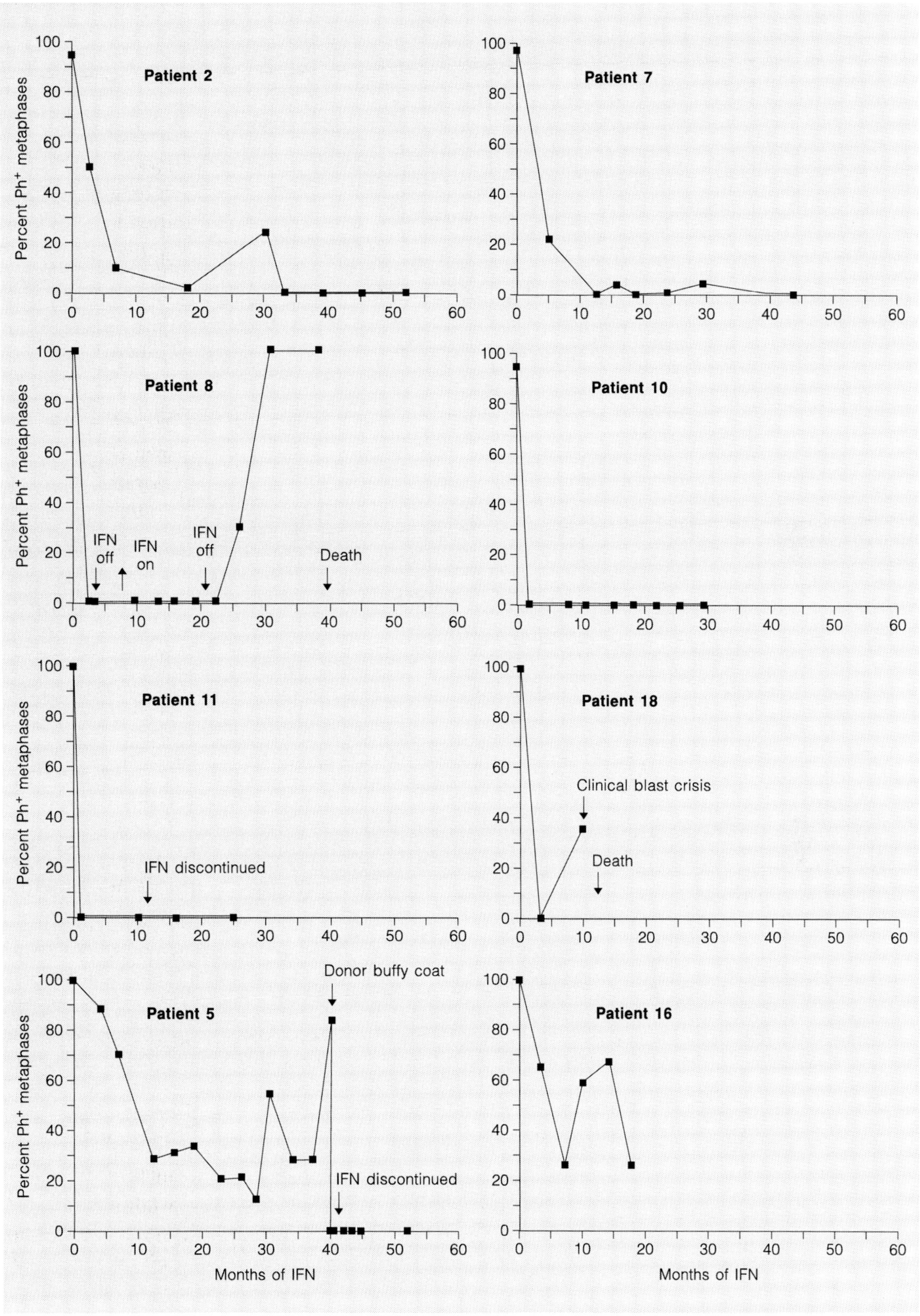

Fig. 1. Time course of cytogenetic response for all responding patients.

Outcome of Nonresponders

Ten patients did not have a cytogenetic response. Four patients developed accelerated phase/blast crisis within the first 3–6 months of IFN treatment. Two of these patients remain alive 14+ and 28+ months after going off study: 1 had a third BMT but is still Ph^+; the other is receiving hospice care at home. The other 2 patients died in blast crisis 9 and 21 months after stopping IFN.

The remaining 6 patients were treated for 9–13 months without a significant decline in the percent of Ph^+ metaphases and were taken off study. One patient went on to second BMT at the time of acceleration and died of transplant-related complications on day 10; 1 patient was maintained on hydroxyurea for 29 months and died in blast crisis; 1 patient was maintained on hydroxyurea and anagrelide and had a single donor buffy coat infusion without response, 1 patient continues on IFN 3 times per week and hydroxyurea, 1 patient remains on IFN for hematological control, and 1 patient's fate is unknown.

Dose Escalation and Maintenance

The planned dose escalation was not tolerated by most patients. The median tolerated induction dose was 3.1×10^6 IU/m^2/day (range $1–6.0 \times 10^6$). There was no difference in median induction doses between those who did and did not have a cytogenetic response.

Five of the 6 patients who achieved a complete cytogenetic response have been on treatment for over 1 year. All patients have required a change in the dose schedule (i.e., from 3 to 5 times per week or q.o.d.) and some also required an increase in dose to maintain a complete cytogenetic response.

Side Effects

Most patients experienced a transient flu-like syndrome after starting IFN. Persistent low grade fatigue was common with long-term therapy, however, working patients were able to maintain full-time jobs. Significant toxicity consisting of profound fatigue, weight loss, nausea and vomiting, diarrhea, confusion, or depression occurred in 6 patients. Five of the 6 patients did not have a cytogenetic response to IFN. The 6th patient, a responder, had cellulitis at the IFN injection sites and lost 14 kg that he regained after the infection was treated. One patient who started IFN on day 40 after BMT developed chronic GvHD on day 72 and IFN was ultimately discontinued after 1 year in cytogenetic remission. One patient had a transient loss of vision in one eye. Complete neurological workup including head MRI and spinal fluid exams was normal. A similar TIA-like episode occurred at a later time while on IFN but not since IFN was stopped 42+ months ago.

Discussion

This is the first study that prospectively examines the use of IFN to treat relapse of CML in chronic phase after BMT. Previous case reports [11–14] have also documented that IFN can control blood counts and can occasionally produce complete cytogenetic responses. Most of the patients in these reports had relapsed after receiving T-cell-depleted marrow and were treated with IFN thrice weekly. This study differs from these reports in that all patients had hematological as well as cytogenetic relapse, all but 1 patient relapsed after receiving non-T-cell-depleted marrow (for first BMT), and all patients received maximum tolerated daily doses of IFN for induction.

IFN treatment of patients who relapse with chronic phase CML after BMT is feasible and effective. Just as in those with newly diagnosed CML treated with IFN, abnormal blood counts are controlled in 75% of patients. The proportion of post-transplant patients who achieve complete cytogenetic response (33%) appears to be at least equivalent to that (19%) reported for patients with newly diagnosed CML. The time to achieve a cytogenetic response is shorter and the doses of IFN required are lower than in nontransplant patients.

Whereas IFN appears to be effective in de novo treatment of CML because of its antiproliferative effects on CML progenitor cells, there may be other mechanisms responsible for the effects seen in the post-transplant patient. IFN increases the expression of the major histocompatibility antigens that may facilitate donor lymphocyte recognition and suppression of leukemic host cells. In addition, IFN appears to increase the number of natural killer cells, some of which may have lymphokine-activated killer activity with antileukemic effect after allografting [17].

Although acquisition of a new clonal cytogenetic abnormality is considered predictive of the development of accelerated phase in CML patients who have not undergone BMT, the presence of additional clones in the post-BMT setting did not have the same significance. New clonal rearrangements were present in the majority of patients studied, including 4 patients with a complete cytogenetic response and 1 with partial cytogenetic response. Most of these clones are likely due to chromosome damage resulting from the conditioning regimen. Clonal evolution of CML indicative of acceleration, particularly in the case of trisomy 8, however, cannot be ruled out. In either event, the identification of additional clones in a patient who is otherwise clinically in chronic phase does not appear to predict for nonresponse to IFN or to the development of accelerated phase when patients are treated with IFN.

Most post-transplant patients did not tolerate dose escalation beyond 3×10^6 IU/m^2/day, generally because of myelosuppression. In the post-transplant setting dose schedule appears to be important. The thrice weekly maintenance schedule was not adequate to maintain a cytogenetic remission. Differences in dose schedule may account for lower cytogenetic response rates reported by others who have retrospectively examined the use of IFN to treat post-BMT relapse [12, 14].

Only 1 patient experienced significant worsening of underlying GvHD. Since this patient started IFN on day 40 after transplant, it is unclear whether IFN played a role in the development of GvHD. This patient achieved a complete cytogenetic response while 5 other patients with complete cytogenetic response have no chronic GvHD.

IFN is effective in treating post-transplant relapse of CML in chronic phase, however, its impact on survival remains unclear and long-term follow-up will be required. This study suggests that there may be a role for the use of IFN in the immediate post-transplant setting in high-risk patients or for patients with early cytogenetic relapse only. To date, 12 patients who have had cytogenetic relapse only have been treated with IFN in a similar manner. Of 9 patients evaluable for cytogenetic response, 7 have achieved a complete cytogenetic response, 1 a partial cytogenetic response, 1 had no response, developed blast crisis and died, and one patient died early on study. These early results are encouraging but will require more patients and longer follow-up before any conclusions can be made.

Acknowledgements

This study could not have been done without the assistance of Karen Lee and the support of Dr. Judith Prestifilippo of Hoffmann-La Roche. Special thanks to Kathryn Morgan, who performed most of the cytogenetic studies, and to Dr. Abraham Grossman for the Southern blot studies.

We would like to thank and acknowledge the physicians who referred and cared for the patients on this study: Saramma Alexander, George Atweh, Byron Chesbro, Richard Dobrow, Donald Doll, Robert Gersh, Barbara Grant, Gerhard Johnson, Kevin Kane, John Kessler, David Kwiatkowski, Judith Lyding, Gordon Nakano, Margaret O'Donnell, Yahia Rahim, Aaron Rausen, Jeffrey Wisch.

References

1 Goldman JM, Apperley JF, Jones L, Marcus R, Goolden AWG, Batchelor R, Hale G, Waldmann H, Reid CD, Hows J, Gordon-Smith E, Catovsky D, Galton DAG: Bone marrow transplantation for patients with chronic myeloid leukemia. N Engl J Med 1986;314:202.

2 Thomas ED, Clift RA, Fefer A, Appelbaum FR, Beatty P, Bensinger WI, Buckner CD, Cheever MA, Deeg HJ, Doney K, Flournoy N, Greenberg P, Hansen JA, Martin P, McGuffin R, Ramberg R, Sanders JE, Singer JW, Stewart P, Storb R, Sullivan K, Weiden PL, Witherspoon R: Marrow transplantation for the treatment of chronic myelogenous leukemia. Ann Intern Med 1986;104:155.

3 Sanders JE, Buckner CD, Clift RA, Fefer A, McGuffin R, Storb R, Appelbaum F, Bensinger W, Beatty P, Doney K, Durnam D, Martin P, Sullivan K, Stewart P, Witherspoon RP, Thomas ED: Second marrow transplants in patients with leukemia who relapse after allogeneic marrow transplantation. Bone Marrow Transplant 1988;3:11.

4 Talpaz M, Kantarjian H, Kurzrock R, Trujillo JM, Gutterman JU: Interferon alpha produces sustained cytogenetic responses in chronic myelogenous leukemia Philadelphia chromosome-positive patients. Ann Intern Med 1991; 114:532.

5 Grossman A, Silver RT, Arlin Z, Colman M, Camposano E, Gascon P, Benn PA: Fine mapping of chromosome 22 breakpoints within the breakpoint cluster region (bcr) implies a role for bcr exon 3 in determining disease duration in chronic myeloid leukemia. Am J Hum Genet 1989;45:729.

6 Durnam DM, Anders KR, Fisher L, O'Quigley J, Bryant EM, Thomas ED: Analysis of the origin of marrow cells in bone marrow transplantation recipients using a Y chromosome specific in situ hybridization assay. Blood 1989;72:2220.

7 Gehly GB, Bryant EM, Lee AM, Kidd PG, Thomas ED: Chimeric BCR-abl messenger RNA as a marker for minimal residual disease in patients transplanted for Philadelphia chromosome-positive acute lymphoblastic leukemia. Blood 1991;78:458.

8 Kawasaki ES, Clark SS, Coyne MY, Smith CD, Champlin R, Witte ON, McCormick FP: Diagnosis of chronic myeloid and acute lymphocytic leukemias by detection of leukemia-specific mRNA sequences amplified in vitro. Proc Natl Acad Sci USA 1988;85:5698.

9 Talpaz M, Kantarjian HM, McCredie KB, Trujillo JM, Keating MJ, Gutterman JU: Hematologic remission and cytogenetic improvement induced by human interferon alpha in chronic myelogenous leukemia. N Engl J Med 1986; 314:1065.

10 Arthur CK, Apperley JF, Guo AP, Rassool F, Gao LM, Goldman JM: Cytogenetic events after bone marrow transplantation for chronic myeloid leukemia in chronic phase. Blood 1988;71:1179.

11 Newland AC, Jones L, Mir N, Slocombe GW: Alpha 2 interferon in chronic myeloid leukemia following relapse post-allogeneic transplant. Br J Haematol 1987;66:141.

12 Arcese W, Mauro FR, Almena G, Screnci M, DeCuia MR, Iori AP, LoCoco F, Fazi P, Cedrone M, Mandelli F: Interferon therapy for Ph positive CML patients relapsed after T cell depleted allogeneic bone marrow transplantation (abstract). New Strategies in Bone Marrow Transplantation, UCLA Symp, 1990, p 281.
13 Alimena G, De Cuia MR, Mecuccia C, Arcese W, Mauro F, Screnci M, Mancini M, Cedrone M, Nanni M, Montefusco E, Mandelli F: Cytogenetic follow-up after allogeneic bone marrow transplantation for Ph positive chronic myelogenous leukemia. Bone Marrow Transplant 1990;5:119.
14 Calabrese G, Di Bartolomeo P, Stuppia L, Franchi PG, Parruti G, Ciancarelli M, Angrilli F, Geraci L, Palka G: Cytogenetics in patients with chronic myelogenous leukemia treated with bone marrow transplantation. Cancer Genet Cytogenet 1989;41:49.
15 Weiden PL, Sullivan KM, Flournoy N, Storb R, Thomas ED, Seattle Bone Marrow Transplant Team: Antileukemic effects of chronic graft versus host disease: Contribution to improved survival after allogeneic transplantation. N Engl J Med 1981;304:1529.
16 Horowitz MM, Gale RP, Sondel PM, Goldman JM, Kersey J, Kolb HJ, Rimm AA, Ringden O, Rozman C, Speck B, Truitt RL, Zwaan FE, Bortin MM: Graft versus leukemia reactions after bone marrow transplantation. Blood 1990;75:555.
17 Mackinnon S, Bungey J, Chase A, Paulsen W, Hows JM, Goldman JM: Origin and function of adherent lymphokine activated killer cells in patients with chronic myeloid leukemia who relapse following bone marrow transplantation. Br J Haematol 1991;77:60.

Acta Haematol 1993;89(suppl 1):15–21

C.K. Arthur
D.D.F. Ma

Haematology Department,
Royal North Shore Hospital,
Sydney, Australia

Combined Interferon Alfa-2a and Cytosine Arabinoside as First-Line Treatment for Chronic Myeloid Leukemia

Key Words
Cytosine arabinoside
Chronic myeloid leukemia
Interferon
Philadelphia chromosome

Abstract
Thirty patients with chronic myeloid leukemia in chronic phase and less than 1 year from diagnosis were treated with a combination of interferon alfa-2a (IFN) 9 million units daily continuously and intermittent low-dose cytosine arabinoside (Ara-C) 20 mg/m^2 daily for 21 days every 42 days. The leukemia was controlled initially with hydroxyurea prior to commencing IFN and Ara-C. The treatment was continued for at least 12 months after which time nonresponders were withdrawn from the trial and responders continued on IFN alone. The median duration of follow-up is 14 months (range 10–53 months). Hematological response was assessed by clinical and laboratory parameters and cytogenetic response was assessed by regular bone marrow chromosome analysis. A complete hematological response occurred in 28/30 patients (93%). A complete cytogenetic response (no detectable Philadelphia chromosome-positive metaphases) was present on at least one occasion in 9/30 (30%), a partial cytogenetic response (between 1 and 34% Philadelphia chromosome-positive metaphases) in 7/30 (23%) and a minor response in 4/30 (13%), giving an overall cytogenetic response rate of 67%. Significant side effects included mucositis, nausea, cytopenia and depression. Side effects could be managed by dose reduction or temporary cessation and were tolerable in most patients, but in 1 patient this led to withdrawal from the trial due to severe depression. Two patients have transformed, 1 to acute lymphoblastic leukemia and 1 to accelerated phase. Two patients have died after exiting the study, both from complications of allogeneic bone marrow transplantation. In conclusion, these results are superior to the results using IFN alone and indicate the need for a randomized study.

Introduction

Chronic myeloid leukemia (CML) is a leukemic disorder which progresses through several phases, initially a chronic phase and then transforming into acute leukemia after a variable time period [1–3]. There is frequently an intermediate accelerated phase characterized by relative resistance to therapy, more rapid proliferation of the leukemic cells, an increase in immature leukemic cells and often additional cytogenetic abnormalities. CML is characterized by a consistent and unique chromosomal abnormality, the Philadelphia (Ph) chromosome [4]. The Ph chromosome

C.K. Arthur
Haematology Department
Royal North Shore Hospital
St. Leonards
Sydney 2065 (Australia)

0001–5792/93/0895–0015
$ 2.75/0

involves a reciprocal translocation between the long arms of chromosomes 9 and 22 [5]. The translocation results in juxtaposition of the Abelson proto-oncogene on chromosome 9 and a gene known as BCR on chromosome 22. A unique hybrid BCR-ABL fusion gene is produced and a unique protein product with enhanced tyrosine kinase activity is formed [6].

The chronic phase of the disease can be controlled with oral cytotoxic agents such as hydroxyurea and busulfan. The accelerated phase however is more difficult to control and the acute or blastic phase is highly resistant to treatment and at best only transient responses can be achieved. It is transformation to the advanced phases which results in shortening of the life span of patients with CML and an overall median survival of 3–4 years. Conventional treatment with agents like hydroxyurea or busulfan do not seem to prevent the transformation from chronic phase [7].

The factors leading to transformation of the chronic phase are thought to involve new genetic mutations and this is supported by the finding of new chromosomal abnormalities in advanced disease. Certain cytogenetic abnormalities are typically seen and include a duplication of the Ph chromosome, isochromosome 17q and trisomy 8 [8]. Molecular changes include mutations in the p53 tumor suppressor gene and N-ras mutations [9, 10]. The mechanism by which new mutations occur is unknown but the events appear to occur at random. If new mutations are a random event then it is logical to propose that any treatment which reduces the total number of leukemic cells will also reduce the transformation rate. This concept is depicted graphically in figures 1–3. Conventional treatment with hydroxyurea or busulfan will reduce the leukemic cell mass, but the degree of reduction is limited since bone marrow cytogenetic examination even at the time of normalized peripheral blood counts usually shows predominantly or totally Ph-positive metaphases (fig. 2). Hence a true remission or cytogenetic remission of the leukemia is rarely if ever achieved with such therapy [7].

It is almost axiomatic that in order to achieve durable responses in malignant disorders the first step must be achievement of a complete remission. This goal has been consistently achieved in CML through high-dose chemoradiotherapy followed by allogeneic bone marrow transplantation [11, 12]. Allogeneic bone marrow transplantation is however limited to those under age 50 with an HLA-compatible donor. Intensive chemotherapy can achieve cytogenetic remission in over 30% of CML patients [13, 14] and some studies do report improved survival, especially in the poor prognosis subgroup of chronic phase patients [15]. However, the duration of cytogenetic remission with intensive chemotherapy is short. In 1986 Talpaz et al. [16] reported cytogenetic responses in patients treated with alpha interferon (IFN). A recent update of the MD Anderson results from 96 consecutive patients in early chronic phase showed a complete cytogenetic response in 18/96 (19%) [17]. Eleven patients have durable ongoing complete cytogenetic responses from 6 to more than 45 months (median more than 30 months). A comparison with historical controls from the same institution suggested that IFN treatment significantly reduced the transformation rate and the survival of the IFN-treated group was improved (median 62 months versus 39 months in 303 patients at the same institution treated with various chemotherapies). The ability of IFN to induce cytogenetic remission has been confirmed in other studies [18–21], but the question of improved survival remains to be answered in randomized studies. It is logical, however, that the more profound reduction in the leukemic population following IFN could reduce the number of random mutations and therefore reduce the risk of transformation to the acute phase (fig. 3).

The mechanism of action of IFN against CML is uncertain but there are several possibilities [22, 23]. IFN may be acting as an antiproliferative agent. IFN may affect the expression of oncogenes. Enhancement of immune function, in particular natural killer cell activation, may result in antileukemic activity. Finally, IFN has recently been shown to modulate the expression of adhesion molecules on the CML progenitor cells and in vitro IFN restores the impaired binding of CML progenitors to stromal cells and possibly restores control of CML cell growth by the hemopoietic microenvironment [24, 25]. There is some evidence that IFN has a better effect when the underlying disorder is at a state of minimal residual disease, e.g. when in remission rather than in overt disease. In this latter scenario IFN may work best in CML after remission induction with chemotherapy [26].

IFN induces major cytogenetic responses in the minority of patients and so additional methods are needed to improve the response rate. Cytosine arabinoside (Ara-C) has in vitro [27] and in vivo [28] activity against CML. Interestingly, Ara-C can also induce a cytogenetic remission in CML [28], but the mechanism is not known. Animal studies and in vitro models have indicated synergy between IFN and Ara-C [29] and so it is logical to test these agents in combination against CML. Preliminary clinical experience suggests that the combination of IFN and Ara-C improves the results of IFN treatment alone in CML patients [30–32]. This report summarizes our experience with combined IFN and Ara-C in early-phase CML.

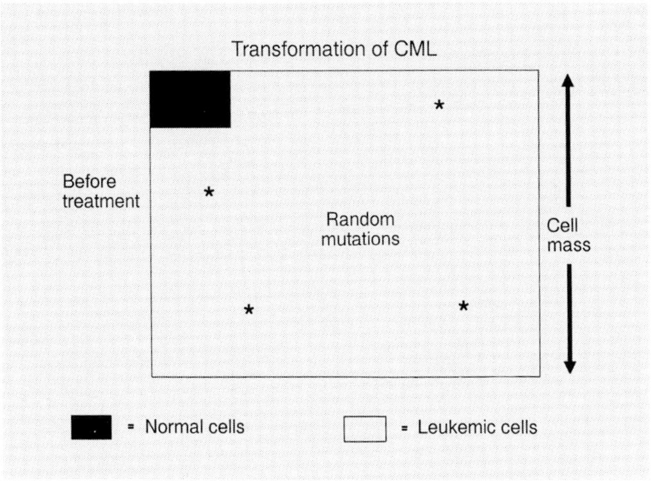

1

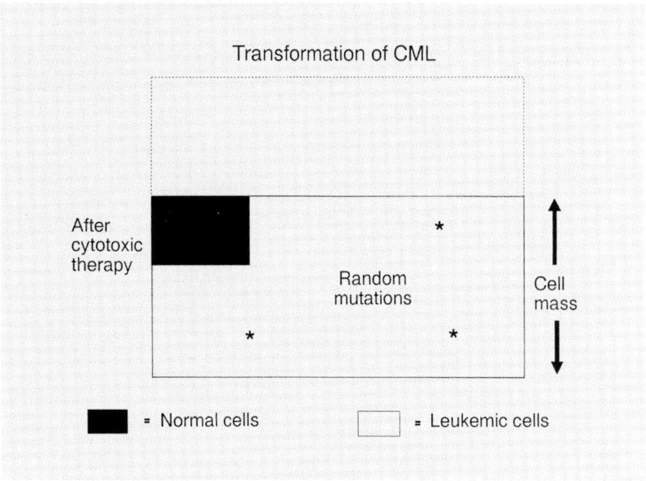

2

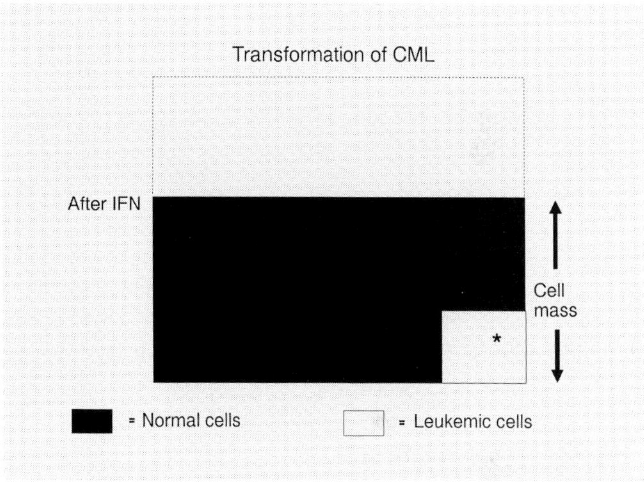

3

Fig. 1. Schematic representation of the hemopoietic progenitor cell mass at the time of diagnosis. A small percentage of normal (Ph-negative) cells are present at the time of diagnosis. The leukemic CML progenitor cells are at risk of random mutational events, which may result in transformation to the acute phase of the disease.

Fig. 2. The mass of hemopoietic progenitors has been reduced by chemotherapeutic agents such as hydroxyurea but the proportion of normal to leukemia is predominantly leukemia. Although the number of mutational events may be reduced the degree of reduction in leukemia is insufficient to result in obvious reduction in transformation risk.

Fig. 3. The mass of hemopoietic progenitors has been reduced by IFN but in contrast to the situation in figure 2 the proportion of normal to leukemic cells is predominantly normal, i.e. cytogenetic remission. Since there has been a more profound reduction of the leukemic progenitors it would follow that there will be a significant reduction in the random mutations and therefore a reduction in the risk of transformation.

Patients and Methods

The study was designed as an open study to test the efficacy and tolerance of IFN in combination with low-dose Ara-C in patients with early-phase CML. Thirty patients from eight participating centers were entered onto the study between October 1987 and May 1991. Cytogenetic analysis was performed on 40 marrow metaphases where possible. Patient characteristics are listed in table 1. The analysis was carried out at 23/6/92.

Inclusion Criteria

Ph-positive CML, age < 65 years, CML in chronic phase (according to criteria of the international bone marrow transplant registry), diagnosis of CML less than 1 year.

Exclusion Criteria

Karnofsky score less than 80, pregnant or lactating women, previous interferon therapy, previous cytotoxic therapy (except in the case of busulfan if less than 7 days duration or hydroxyurea for less than 12 months), significant disease affecting renal, hepatic, cardiac or respiratory systems, leukocytes < 2.0×10^9/l, platelets < 100×10^9/l, surgery or radiotherapy within 1 month prior to study.

Table 1. Patient characteristics (median and range)

Number	30	
Age	44 years	(6–63)
M:F	2.75:1	
Spleen	3 cm	(0–20)
Hb	115 g/l	(64–150)
WCC	147×10^9/l	(8.6–476)
Plt	410×10^9/l	(180–1,908)
Blasts	0%	(0–6)
Sokal index		
<0.8 (good prognosis)	14	(47%)
0.8–1.2 (intermediate prognosis)	12	(40%)
>1.2 (poor prognosis)	4	(13%)

Table 2. Side effects

Side effect	Frequency	Comments
Thrombocytopenia ($<150\times 10/l$)	83%	1 Pt ceased Rx (temporary)
Neutropenia ($<1.0\times 10/l$)	50%	1 Pt ceased Rx (temporary)
Mucositis (Ara-C)	53%	1 Pt ceased Rx (temporary)
Vomiting	23%	1 Pt ceased Rx (temporary)
Depression	20%	1 Pt ceased Rx (permanent)
Anemia requiring transfusion	3 Pt	
Nausea	56%	
Lethargy	46%	
Flu-like symptoms	30%	
Diarrhea	20%	
Skin rash	20%	
Febrile	20%	
Weight loss	16%	
Hypothyroidism	1 Pt	
Pernicious anemia	1 Pt	
Gynecomastia	1 Pt	

Rx = Treatment; Pt = patient.

Treatment

After study entry therapy was commenced with hydroxyurea, initially 2 g daily and then the dose adjusted until the total white cell count (WCC) was $<11\times 10^9$/l. At this time recombinant human interferon alfa-2a (Roferon-A) was introduced at 3 million units daily by deep subcutaneous injection. The dose was escalated by 3 million units every 3 days up to 9 million units daily. IFN was then continued daily for at least 6 months. Once the patient was stabilized on the maximum tolerated dose of IFN or after 4 weeks (whichever was the shorter) then Ara-C was commenced. Ara-C was started at 20 mg/m^2 once daily by deep subcutaneous injection for 21 days and this course was repeated every 42 days. If the WCC rose above 4×10^9/l then hydroxyurea was added and the dose adjusted as necessary until the WCC returned to a level between 1×10^9/l and 4×10^9/l. Toxicity due to IFN or Ara-C was managed by temporary cessation or dose reduction and once the problem resolved an attempt was made to escalate the dose or reintroduce the drug.

Assessment of Response

Hematological response was defined as follows: Complete response is normalization of abnormal clinical signs including palpable splenomegaly and a return of the blood count and white cell differential to normal. A partial response is a fall in the WCC to less than 50% of the pretreatment value and to at least below 20×10^9/l. Cytogenetic response was defined as an increase in the percentage of Ph-negative cells in the marrow. The criteria of Talpaz [16] are used: (1) complete cytogenetic response (CCR) if Ph-positive metaphases are $<1\%$, (2) partial cytogenetic response (PCR) if Ph-positive cells are 1–34%; (3) minor cytogenetic response if Ph-positive cells are 35–95%; (4) no response if Ph-positive cells are $>95\%$. Cytogenetic studies were done at the start of the study, just prior to starting Ara-C and then at 6-weekly interval just prior to each cycle of Ara-C. It was found that analyzable metaphases were more readily obtained if the patient ceased IFN for 2 days before bone marrow examination and this became the usual practice during the course of the study.

Results

The median follow-up for the 30 patients is 14 months (10–53 months). A complete hematological response was obtained in 28/30 patients. Two patients achieved a partial hematological response.

A CCR was achieved in 9/30 (30%) patients and a PCR was obtained in 7/30 (23%) patients. Hence 16/30 (53%) patients achieved a major (complete + partial) cytogenetic response. A further 4/30 (13%) achieved a minimal response. At the time of reporting a CCR is still present in 4/30 (13%) and 7/30 (23%) have a PCR, giving a persisting major cytogenetic response of 37%.

As an indication of the rapidity of the response the median time to reach the first PCR was 8 months (2–12 months) and in the 9 complete responders the median time to reach CCR was 12 months (3–28 months). The median time to reach the best response in all responding patients was 10 months (2–38 months).

Two patients have transformed, one to lymphoid blast crisis 6 months after electively stopping IFN and a second patient transformed to accelerated phase while on IFN. Two patients have died, both from complications of allogeneic bone marrow transplantation.

Side effects of treatment (table 2) were generally tolerable and in only 1 patient was therapy permanently discontinued because of toxicity. The side effects of IFN included influenza-like symptoms and were controlled by paracetamol or dose reduction. Ara-C had additive toxicity with the main side effect being mouth ulceration which occurred 5–10 days after starting Ara-C and was severe (WHO grade > 2) in 3 patients. Nausea was the next most prominent side effect of Ara-C.

Cytopenia was encountered in all patients but it was noted that despite moderate neutropenia and/or thrombocytopenia frequently the marrow remained hypercellular, suggesting that there was significant marrow reserve and dose adjustment according to peripheral blood counts needs to consider this factor. Some unusual side effects included hypothyroidism in 1 patient, pernicious anemia in 1 patient, and gynecomastia in 1.

Discussion

There is now no doubt that IFN can induce cytogenetic response in patients with Ph-positive CML [18–21]. Responding patients appear to have significantly improved median survival, in excess of 5 years, when compared with historical controls [17]. It is possible that IFN simply identifies the subgroup of patients who would have had good survival even without IFN, however, studies looking at all patients on IFN suggest that the survival benefit is still evident when applied to the entire group and not just the responders [17]. Several prospective randomized studies are currently in progress to answer the question of survival.

It is logical that achievement of a cytogenetic remission could reduce the transformation rate in CML [7]. The process of transformation is uncertain but new genetic abnormalities almost certainly play a role. One theory suggests that this is a random event. Any treatment which reduces the mass of leukemic cells at risk of random new mutations should also reduce the transformation rate and hence improve survival. The unique difference between IFN therapy and other cytotoxic therapy such as hydroxyurea is the ability of IFN to induce a selective suppression of the leukemia for a similar degree of overall cytoreduction. There is a further mechanism by which IFN may suppress transformation and this is via the postulated action of IFN on minimal residual disease [26]. In a number of hematological malignancies IFN shows minimal activity in advanced disease but appears beneficial in situations of minimal residual disease after chemotherapy. Some examples include IFN after remission induction in myeloma and large cell lymphoma [33–35]. It is possible therefore that IFN is exerting an effect on transformed clones of CML when they are at a very early stage, and for this mechanism to operate a prior cytogenetic response would not be necessary. Ara-C could also have a beneficial effect at the stage of minimal disease.

This study was initiated on the assumption that improving the cytogenetic response rate might further reduce transformation rate and improve survival. Improved response rates of many malignancies have followed the combination of active single agents. Both IFN and Ara-C used as single agents can induce a cytogenetic response in CML and the combination has demonstrated synergy in vitro [29]. It is therefore logical to combine these agents in the treatment of CML. Our study has demonstrated that the combination of IFN and Ara-C is tolerable, although the addition of Ara-C increases mucositis and cytopenia. The major (CCR + PCR) cytogenetic response in our study was 53%, which compares favorably with the response to IFN alone of approximately 26%. A preliminary report from a French multicenter randomized study comparing IFN alone with IFN plus Ara-C gave almost identical results to our study with 28% CCR and 44% major cytogenetic responses (CCR + PCR) in the IFN + Ara-C arm compared with 8 and 31% in the IFN-only arm [36]. The dose of Ara-C in our study is higher than in the French study and further investigation will be required to define the optimal dose.

Although it will be important to identify if IFN or the combination of IFN and Ara-C leads to a survival advantage there may be other benefits of therapies which can increase the cytogenetic response. One possible benefit may derive through autologous bone marrow transplantation [37, 38]. In vivo purging of CML with IFN and Ara-C could reduce the leukemic cell contamination of autologous marrow and in combination with other strategies improve the results of autologous BMT in CML. Finally, it is a basic premise that the first step in cure of a malignancy requires the attainment of a complete remission. The cure of CML with chemotherapy has not been achieved, excluding of course marrow-ablative chemoradiotherapy and allogeneic BMT. Although there is no evidence that IFN alone or IFN + Ara-C can cure any patients, nevertheless by consistently achieving complete cytogenetic remissions it provides a direction forward in those who cannot benefit from an allogeneic BMT.

Although our results are encouraging there are approximately 50% of patients who fail to achieve a cytogenetic response. There are several strategies that might be tried in these patients if they are not suitable for allogeneic bone marrow transplantation. Such strategies include autologous bone marrow transplantation and intensive combination chemotherapy followed by IFN. It is interesting, however, that dose escalation of IFN has not been explored. It is known that IFN shows a dose response in the range 2–9 million units/day but there is little experience above this dose [18, 21, 39]. Shtalrid et al. [40] reported IFN dose escalation to 12 million units/day in relapsing or nonresponding patients and obtained responses in 2/8 patients. Recently Taylor et al. [4] reported dose escalation to 18 million units/day in a patient who had achieved a CCR on IFN but then developed clonal progression with a double Ph chromosome. The high-dose IFN was followed by a reduction in the double Ph clone and the patient has remained on 18 million units/day for 9 months.

We have initiated a study of high-dose IFN in poor prognosis patients (Sokal Index greater than or equal to 1.2) or patients failing to achieve a cytogenetic response with combined IFN/Ara-C. Our preliminary experience with IFN at a dose of 18 million units/day suggests that this dose is well tolerated. One patient has been treated on the study for 4 months and has had no serious side effects. Two other patients who have been resistant to IFN and Ara-C have received high-dose IFN for 1 and 4 months and this was well tolerated in both patients. One patient has undergone elective autologous marrow transplant and the other continues on high-dose IFN. It is planned to administer corticosteroids to ameliorate side effects of high-dose IFN, but so far this has not been necessary. It is too early to judge if there will be cytogenetic responses to high-dose IFN in these poor prognosis patients.

Acknowledgements

We wish to thank the doctors who participated in this study or referred patients to us: Dr. Michael Stevens (Royal Alexandra Hospital for Children, Camperdown), Dr. Frank Firkin (St. Vincents Hospital, Melbourne), Dr. Michael Seldon (Royal Newcastle Hospital), Dr. Alec Morley (Flinders Medical Centre, Adelaide), Dr. Pamela Harrison (Mater Misericordiae Hospital, Newcastle), Dr. Chris Juttner and Dr. Bik To (Institute of Medical and Veterinary Science, Adelaide), Dr. Harry Iland (Kanematsu Institute, Royal Prince Alfred Hospital, Sydney), Dr. Robert Ravich, Dr. Steve Flecknoe-Brown. This study was supported with a grant from Hoffmann-La Roche.

References

1 Goldman JM: Management of chronic myeloid leukaemia. Scand J Haematol 1986;37: 269–279.

2 Kantarjian HM, Talpaz M, Gutterman JU: Chronic myelogenous leukemia: Past, present and future. Hematol Pathol 1988;2:91–120.

3 Silver RT: Chronic myeloid leukemia: A perspective of the clinical and biological issues of the chronic phase; in Cannelos GP (ed): Chronic Leukemias. Hematology/Oncology Clinics of North America. Pennsylvania, Saunders, 1990, vol 4, pp 319–335.

4 Nowell PC, Hungerford DA: A minute chromosome in human chronic granulocytic leukaemia. Science 1960;132:1497–1501.

5 Rowley JD: A new consistent chromosomal abnormality in chronic myelogenous leukaemia identified by quinacrine fluorescence and Giemsa staining. Nature 1973;243:290–291.

6 Gale RP, Cannani E: The molecular biology of chronic myelogenous leukaemia. Br J Haematol 1985;60:395–408.

7 Sokal JE: Prognosis in chronic myeloid leukaemia: Biology of the disease vs. treatment; in Goldman JM (ed): Chronic Myeloid Leukaemia. Baillière's Clinical Haematology. London, Baillière Tindall, 1987, vol 1, pp 907–930.

8 Hagemeijer A: Chromosome abnormalities in CML; in Goldman JM (ed): Chronic Myeloid Leukaemia. Baillière's Clinical Haematology. London, Baillière Tindall, 1987, vol 1, pp 963–981.

9 Foti A, Ahuja HG, Allen SL, Koduru P, Schuster MW, Schulman P, Bar-Eli M, Cline MJ: Correlation between molecular and clinical events in the evolution of chronic myelocytic leukemia to blast crisis. Blood 1991;77: 2441–2444.

10 Ahuja HG, Foti A, Bar-Eli M, Cline MJ: The pattern of mutational involvement of RAS genes in human hematologic malignancies determined by DNA amplification and direct sequencing. Blood 1990;75:1684.

11 Goldman JM, Apperley JF, Jones L, Marcus R, Goolden AWG, Batchelor R, Hale G, Waldmann H, Reid CD, Hows J, Gordon-Smith E, Catovsky D, Galton DAG: Bone marrow transplantation for chronic myeloid leukaemia. N Engl J Med 1986;314:202–207.

12 Thomas ED, Clift RA, Fefer A, Appelbaum FR, Beatty P, Bensinger WI, Buckner CD, Cheever MA, Deeg HJ, Doney K, Flournoy N, Greenberg P, Hansen JA, Martin P, McGuffin R, Ramberg R, Saunders JE, Singer J, Stewart P, Storb R, Sullivan K, Weiden PL, Witherspoon R: Marrow transplantation for the treatment of chronic myelogenous leukaemia. Ann Intern Med 1986;104:155–163.

13 Kantarjian HM, Vellekoop L, McCredie KB, Keating MJ, Hester J, Smith T, Barlogie B, Trujillo J, Freireich E: Intensive combination chemotherapy (ROAP 10) and splenectomy in the management of chronic granulocytic leukemia. J Clin Oncol 1985;3:192–200.

14 Cunningham I, Gee T, Dowling M, Chaganti R, Bailey R, Hopfan S, Bowden L, Turnbull A, Knapper W, Clarkson B: Results of treatment with Ph^1 chronic myelogenous leukemia with an intensive treatment regimen (L-5 protocol). Blood 1979;53:375–395.

15 Kantarjian HM, Smith TL, McCredie KB, Keating MJ, Walters RS, Talpaz M, Hester JP, Bligham G, Gehan E, Freireich E: Chronic myelogenous leukemia: A multivariate analysis of the associations of patient characteristics and therapy with survival. Blood 1985;66: 1326–1335.

16 Talpaz M, Kantarjian HM, McCredie K, Trujillo JM, Keating MJ, Gutterman JU: Hematologic remission and cytogenetic improvement induced by recombinant human interferon alpha A in chronic myelogenous leukaemia. N Engl J Med 1986;314:1065–1069.

17 Talpaz M, Kantarjian H, Kurzrock R, Trujillo JM, Gutterman JU: Interferon-alpha produces sustained cytogenetic responses in chronic myelogenous leukemia. Ann Intern Med 1991; 114:532–538.

18 Alimena G, Morra E, Lazzarino M, Liberati AM, Montefusco E, Inverardi D, Bernasconi P, Mancini M, Donti E, Grignani F, Bernasconi C, Dianzani F, Mandelli F: Interferon alpha-2b as therapy for Ph^1-positive chronic myelogenous leukemia: A study of 82 patients treated with intermittent or daily administration. Blood 1988;72:642–647.

19 Niederle N, Kloke O, Osieka KR, Wandl U, Opalka B, Schmidt G: Interferon alpha-2b in the treatment of chronic myelogenous leukemia. Semin Oncol 1987;16:29–35.

20 Ozer H, Dear K, Testa J, Arthur D, Cooper R, Pettenati M, Rao K, Peterson BA, Schiffer C, Bloomfield CD, for CALG-B: Prolonged administration of subcutaneous α-interferon induces major clinical and complete cytogenetic remissions in untreated Philadelphia positive (Ph^1+) chronic myelogenous leukemia. Proc Am Soc Clin Oncol 1990, p 203a.

21 Freund M, Von Wussow P, Diedrich H, Eisert R, Link H, Wilk H, Buchholz F, Leblanc S, Fonatsch C, Deicher RH, Poliwoda H: Recombinant human interferon (IFN) alpha-2b in chronic myelogenous leukaemia: Dose dependency of response and frequency of neutralizing antibodies. Br J Haematol 1989;72:350–356.

22 Borden EC, Ball AL: Interferons: Biochemical, cell growth inhibitory, and immunological effects; in Progress in Haematol 1985, pp 299–336.

23 Hersey P: The evolving role of alpha interferon in the treatment of malignancies. Aust NZ J Med 1986;16:425–437.

24 Upadhyaya G, Deisseroth AB, Emerson SG: CML progenitor cells are deficient in cell surface LFA-3: Modulation by alpha-interferon in vitro and in-vivo. Blood 1988;72(suppl 1):836a.

25 Dowding C, Guo AP, Osterholz J, Siczkowski M, Goldman J, Gordon M: Interferon-α overrides the deficient adhesion of chronic myeloid leukemia primitive progenitor cells to bone marrow stromal cells. Blood 1991;78:499–505.

26 Shtalrid M, Lugassy G, Berrebi A: High response rate to recombinant interferon alpha in chronic myeloid leukemia after intensive cytoreductive chemotherapy. Blood 1989;74 (suppl 1):370a.

27 Sokal JE, Gockerman JP, Bigner SH: Evidence for a selective effect of cytosine arabinoside in chronic granulocytic leukaemia. Leuk Res 1988;1:453.

28 Sokal J, Bigner S: Low dose cytosine arabinoside in early and advanced chronic granulocytic leukemia. Blood 1986;66(suppl 1):233a.

29 Imanishi J, Tanaka A: Experimental research from clinical applications of interferon. Jpn J Cancer Chemother 1984;11:53–59.

30 Guilhot F, Tanzer J, Brizard A, Dreyfus B: High rate of complete cytogenetic remissions (CCR) in Ph+ chronic myelocytic leukemia (CML) using interferon alpha 2a (IFN)+ hydroxyurea (HU)+cytosine arabinoside in patients with no prior therapy (6/11) or previously treated with prolonged conventional chemotherapy (3/11). Proc Am Assoc for Cancer Res 1990, p 195.

31 Silver RT, Benn P, Szatrowski TP, Verma R, Gutfriend A: Infusional cytosine arabinoside (ARA-C) and recombinant interferon-α (rIFN-α) for the treatment of chronic myeloid leukemia (CML). Proc Am Soc Clin Oncol 1990, p 209.

32 Kantarjian HM, Keating MJ, Estey EH, O'Brien S, Pierce S, Beran M, Koller C, Feldman E, Talpaz M: Treatment of advanced stages of Philadelphia chromosome-positive chronic myelogenous leukemia with interferon-α and low dose cytarabine. J Clin Oncol 1992;10:772–778.

33 Mandelli F, Avvisati G, Amadori S: Maintenance treatment with recombinant interferon alfa-2b in patients with multiple myeloma responding to conventional induction chemotherapy. N Engl J Med 1990;322:1430–1434.

34 Aviles A, Garcia EL, Guzman R, Talavera A, Diaz-Maqueo JC: Interferon alpha 2b as maintenance therapy for large cell lymphoma. Blood 1990;76(suppl 1):1354a.

35 Ascensao JL, Miller KB, Bergstrom SB, Schenkein JJ, Quinn JJ, Erba J, Dainiak N, Poliquin C: Immunotherapy following autologous marrow transplantation for relapsed lymphomas: The role of interferon alpha 2B. Exp Hematol 1991;19(suppl 1):336a.

36 Guilhot F, Tanzer J: Randomised trial of interferon with or without cytosine arabinoside in treatment of chronic myeloid leukemia. Proc Int Soc Haematol Meet 1991, p 311a.

37 Meloni G, De Fabritiis P, Alimena G, Malagnino F, Montefusco E, Sandrelli A, Pinto R, Vignetti M, Lo Coco F, Mandelli F: Autologous bone marrow or peripheral blood stem cell transplantation for patients with chronic myelogenous leukemia in chronic phase. Bone Marrow Transplant 1989;4(suppl 4):92–94.

38 Alimena G, Morra E, Lazzarino M, Liberati AM, Montefusco E, Inverardi D, Bernasconi P, Mancini M, Donti E, Grignani F, Bernasconi C, Dianzani F, Mandelli F: Interferon alpha-2b as therapy for patients with Ph^1-positive chronic myelogenous leukemia. Eur J Hematol 1990;45(suppl 52):25–28.

39 Gastl G, Aulitzky Q, Tilg H, Huber H, Hausmaninger H, Seewann HL, Coser C, Prinroth P, Huber C: Dose related effectiveness of alpha interferon in chronic myelogenous leukemia. Blut 1987;54:251–252.

40 Shtalrid M, Lugassy G, Berebbi A: Treatment of chronic myeloid leukemia with interferon alpha (Roferon-A): Results of the Israeli Study Group on CML. Proc Int Soc Hematol Meet (European Division) 1992, p 260a.

41 Taylor KMcD, Williams B, Elliot SL, Bunce I, Wright S, Rodwell RL, Taylor D: α-Interferon (α-IFN) dose escalation to suppress clonal evolution in chronic myelogenous leukemia. Proc Haematol Soc Aust 1992, p 130.

Acta Haematol 1993;89(suppl 1):22–27

P. Fenaux

Service des Maladies du Sang,
CHU, Lille, France

The Role of All-*Trans*-Retinoic Acid in the Treatment of Acute Promyelocytic Leukemia

Key Words
Acute promyelocytic leukemia
All-*trans*-retinoic acid

Abstract
All-*trans*-retinoic acid (ATRA) has demonstrated in vivo and in vitro that it could differentiate leukemic cells in acute promyelocytic leukemia (APL). ATRA yields complete remission (CR) rates of approximately 90% in newly diagnosed and first relapsing APL, in APL resistant to one or two courses of chemotherapy, whereas patients in second and subsequent remissions often have only partial remissions. The use of ATRA in APL is associated with two major drawbacks: (1) the risk of rapidly rising leukocyte counts, leading to the 'retinoic acid syndrome' which may be fatal if the increase in leukocytes is not reversed and (2) almost universal relapse, if no intensive chemotherapy is administered after CR achievement. Preliminary results, however, suggest that ATRA followed by intensive chemotherapy has improved the outcome of newly diagnosed APL, by slightly increasing the CR rate but also by reducing the risk of relapse, as compared to chemotherapy alone.

Introduction

Since the landmark paper of Bernard et al. [1], anthracyclines (especially daunorubicin) with or without Ara C have been the mainstay in the treatment of acute promyelocytic leukemia (APL). This treatment yields complete remission (CR) rates of about 75%, and 35–40% of complete remitters have prolonged CR with adequate consolidation and maintenance chemotherapy [2–8]. Failure to achieve CR results mainly from death due to exacerbation of coagulation disorders during treatment [9], or sepsis during the chemotherapy-induced period of aplasia rather than true leukemic resistance, which is rare in APL [1–8].

In vitro studies have shown that all-*trans*-retinoic acid (ATRA) was capable of differentiation of APL promyelocytes at relatively low concentrations [10, 11]. This property has been applied in vivo as first-line therapy or in relapsing APL with the hope that, because ATRA did not increase coagulation disorders and did not lead to bone marrow aplasia, it might increase the CR rate. One could also hope that the combination of ATRA and chemotherapy might further reduce the leukemic burden in APL, and thus diminish the risk of relapse and improve the cure rate. This short review will be restricted to clinical results obtained with ATRA and future directions which can be envisaged with this drug in APL.

Pierre Fenaux
Service des Maladies du Sang, CHU
1, Place de Verdun
F–59037 Lille (France)

0001–5792/93/0895–0022
$ 2.75/0

Clinical Results Published with ATRA in APL

ATRA Gives a High CR Rate in APL

High CR rates are obtained, both in newly diagnosed APL and in APL relapsing after prior chemotherapy. In newly diagnosed APL, Huang et al. [11] obtained CR in 15/16 treated patients, Chen et al. [12] in 41/44 patients, Frankel et al. [13] in 27/32, and Fenaux et al. [14] in 25/26 patients. In first relapsing APL, Degos et al. [15] reported CR in 26/28 treated patients, and Warrell et al. [16] in 18/20 treated patients. In second or subsequent relapse, however, results are less favorable: 1 CR in 5 patients reported by Castaigne et al. [17].

In newly diagnosed or first-relapse APL, the very small number of failures to reach CR with ATRA were generally due to early death rather than to leukemic resistance. Early death usually resulted from leukostasis or to the 'retinoic acid syndrome' and less often from bleeding diathesis. In addition, several of the patients who died from bleeding already had CNS bleeding when ATRA was started. Resistance to ATRA, by contrast, was exceptional (2 or 3 patients treated in first relapse or as first-line therapy).

ATRA Is Active through Differentiation of APL Cells and Not through Cell Destruction

This has been clearly shown by: (1) morphological assessment of bone marrow cells during treatment, showing the presence of Auer rods in neutrophils, and the absence of any phase of bone marrow aplasia [17]; (2) immunological studies, demonstrating the presence of an intermediate population of CD_{33}+ CD_{15}+ cells [17]; (3) fluorescence in situ hybridization studies, showing the presence of t(15;17) in maturing cells [18], and (4) clonality studies, showing that circulating neutrophils observed after a few days of ATRA were clonal [19, 20].

In addition to the disappearance of APL cells and progressive normalization of blood count, remission obtained with ATRA was eventually complete, with the subsequent disappearance of t(15;17) translocation [18] and reappearance of polyclonal neutrophils [19, 20]. This suggested that CR was obtained by differentiation of abnormal cells into mature neutrophils which eventually died and were replaced by polyclonal hemopoiesis. CR was usually obtained after 30–80 days.

Importantly, Huang et al. [11] and Chomienne et al. [10] found a close correlation between in vivo and in vitro differentiation of APL cells with ATRA. The rare cases of APL showing no or little differentiation with ATRA in vitro corresponded to patients treated in second or subsequent relapse, or relapsing under ATRA, and who did not respond to ATRA in vivo [17]. In other situations (newly diagnosed or first relapsing APL), the absence of in vitro differentiation of blasts with ATRA strongly argues against APL and should lead to reconsidering the diagnosis (with reassessment of morphological and cytogenetic findings). In these cases, the final diagnosis is usually another type of acute myeloid leukemia, where differentiation of blasts with ATRA is rarely seen, and when present, is always incomplete [10].

ATRA Leads to Early Improvement of Coagulation Parameters

Although the coagulation disorder leading to bleeding diathesis in APL is generally attributed to disseminated intravascular coagulation, several findings suggest that fibrinogenolysis and even more generalized proteolysis may be present [9]. With chemotherapy, a transient increase in coagulation disorders is often observed, even in patients receiving concomitantly heparin or antifibrinolytic agents. By contrast, ATRA leads within 48 h of treatment to an increase in fibrinogen levels, and a decrease in fibrin degradation products and *D*-dimers. More precise analysis of coagulation parameters suggests that ATRA is more effective on fibrinogenolysis/proteolysis than on disseminated intravascular coagulation [21]. Clinically, rapid improvement of coagulation parameters in APL patients treated with ATRA has led to a reduction of early deaths due to bleeding (to about 2% in our recent experience).

Furthermore, the risk of death due to bleeding seems to be higher in APL patients presenting with hyperleukocytosis, when chemotherapy is started. Our preliminary results [Dombret et al., unpublished] suggest that combining chemotherapy and ATRA, in those patients, stabilizes coagulation disorders and reduces bleeding diathesis.

ATRA Alone Cannot Cure APL

In newly diagnosed APL, patients who have reached CR with ATRA alone relapse after a median of 5 months if ATRA is maintained alone [11, 21]. Likewise, most patients relapse within 1 year of CR achievement if low-dose chemotherapy (such as mercaptopurine and methotrexate, or low-dose Ara-C) is administered. The combination of ATRA and low-dose chemotherapy for maintenance may slightly prolong remissions [21]. However, patients relapsing while on ATRA (with or without chemotherapy) fail to reobtain CR with ATRA and eventually have shorter survival [12].

Relapse in patients maintained on ATRA is due to the occurrence of progressive resistance to the drug, and also probably to the fact that ATRA is unable to eradicate clonogenic leukemic cells, because it may be only active on more differentiated cells. Progressive resistance to ATRA in vivo is often but not always correlated with a reduction of in vitro differentiation of APL cells with ATRA [10]. Two possible explanations for this progressive resistance to ATRA have been suggested: pharmacokinetic studies demonstrate a progressive reduction in plasma drug concentrations, which could result from accelerated metabolism of the drug by a cytochrome P-450-like enzyme system [23, 24]; alternatively the induction, by ATRA, of synthesis of cytoplasmic ATRA-binding protein in APL cells could act as a 'buffer' and prevent ATRA from acting in the cell nucleus [25].

Patients in first relapse after chemotherapy who obtain CR with ATRA unfortunately also relapse again, usually within 1 year of achievement of this new CR [15]. This is seen in our experience irrespective of the type of maintenance therapy: ATRA alone, low-dose chemotherapy (such as low-dose Ara-C, or the combination of mercaptopurine and methotrexate), or somewhat more intensive chemotherapy, leading to aplasia (mitoxantrone-Ara-C reinduction courses). By contrast, patients allografted in second CR after ATRA may be cured, although their number in the French experience is too small to draw any significant conclusions. Whether autologous bone marrow transplantation (BMT) or very intensive consolidation chemotherapy can eradicate the disease in APL patients in second CR after treatment with ATRA cannot as yet be determined. In the case of autologous BMT, unless first-remission bone marrow is available, the bone marrow harvest should probably be made only after consolidation chemotherapy, as suggested by a preliminary Italian experience [personal commun.].

Side Effects of ATRA Are Usually Moderate

Dryness of skin and mucosae is almost universally seen at the usual dose of 45 mg/m^2/day of ATRA, but it can be alleviated by cold cream, eyedrops and nasal drops. Bone pain, which was frequently seen in our experience with ATRA manufactured in China, is very rare with the Hoffmann-La Roche drug. Hypertransaminemia and hypertriglyceridemia are seen in $\frac{1}{4}$ and $\frac{1}{2}$ of cases, respectively. Headache is generally moderate in adults. In children, however, intracranial hypertension, with pseudotumor cerebri, has been observed in some cases.

With more widespread use of ATRA, however, very rare but more important side effects have been reported: cutaneous rashes in a few patients [26], erythema nodosum (at least in 3 patients), hypercalcemia (in 2 patients) [26, 27], thromboembolic events (in 1 patient) [28], hyperhistaminemia leading to shock (in 1 patient) [29].

Hyperleukocytosis and the 'Retinoic Acid Syndrome' Are the Major Side Effects of ATRA Therapy

Moderate and transient hyperleukocytosis (peaking at 10,000–15,000/mm^3 around day 15–20 of treatment) is extremely common in APL [11, 17]. In some patients, however, hyperleukocytosis may develop very rapidly, leading to clinical signs of leukostasis and possible fatal outcome. In our early experience with ATRA in newly diagnosed APL in elderly patients (who had a contraindication to anthracycline-Ara-C chemotherapy), approximately one third showed rapidly increasing leukocyte counts and signs of leukostasis [30]. Most of those patients died from probable complications of leukostasis (including stroke, myocardial infarction, pulmonary or renal failure), although their poor general condition certainly also contributed to fatal outcome. Mild chemotherapy (including low-dose Ara-C and hydroxyurea) proved insufficient in stopping the increase in leukocytes. We also found that patients who subsequently developed leukostasis had leukocyte counts above 6,000/mm^3 by day 5 of ATRA treatment, or above 10,000/mm^3 by day 10, or above 15,000/mm^3 by day 15. In subsequent patients who rapidly developed hyperleukocytosis, more intensive chemotherapy (especially daunorubicin-Ara-C combinations) achieved a rapid reduction in leukocyte counts and prevented clinical complications.

More recently, Frankel et al. [13] reported a 'retinoic acid syndrome' (RA syndrome) in 9 of 32 newly diagnosed cases of APL treated with ATRA. The RA syndrome was usually but not always associated to hyperleukocytosis. It consisted of fever, pulmonary infiltrates, pleural and sometimes pericardial effusion, renal impairment and, in some patients, cardiac failure, and usually occurred after 8–15 days of treatment. Leukaphereses proved inconstantly useful in reversing this syndrome, whereas intravenous dexamethasone (20 mg/day), used in 4 patients, led to rapid diminution of symptoms, suggesting a role for the release of cytokines in the pathogenesis of the RA syndrome. Outcome was however fatal in 3 cases.

By retrospectively reviewing our cases, we also found that several of them (including some who did not develop hyperleukocytosis) had indeed presented several signs suggestive of the RA syndrome (especially unexplained fever, pleural or peritoneal effusion, and cardiac failure in 2 patients). Since we decided to start chemotherapy routinely in patients treated with ATRA, as soon as hyperleukocytosis developed, this complication has been rare.

New Approaches to ATRA Treatment in APL

As seen in the previous paragraph, ATRA is capable of inducing CR in the great majority of APL patients. However, its major side effects (hyperleukocytosis and the RA syndrome) need to be prevented. In addition, once CR has been achieved, most patients relapse unless some form of intensive chemotherapy (and/or BMT) is administered. One therefore has to select some form of adjuvant therapy aimed at eradicating the disease. Another goal is trying to overcome the progressive resistance of APL cells to ATRA.

Preventing Hyperleukocytosis and the RA Syndrome

Our early experience suggested that patients whose leukocyte counts were above 6,000/mm^3 by day 5, or 10,000/mm^3 by day 10, or 15,000/mm^3 by day 15 of ATRA treatment were at risk of leukostasis. This prompted us to start a pilot phase of ATRA followed by chemotherapy in newly diagnosed APL. In those patients, a combination of daunorubicin and Ara-C (4+7 course) was added to ATRA in emergency if leukocytes rose above the thresholds defined (>6,000/mm^3 by day 5 or >10,000/mm^3 by day 10, or >15,000/mm^3 by day 15) [14]. Of the 26 patients, 25 achieved CR. In 11 of them, chemotherapy was started in emergency. Criteria for starting chemotherapy were extremely stringent in this study, and possibly resulted in starting chemotherapy in patients who would not have otherwise developed any complication of hyperleukocytosis. However, this attitude, reproduced in our current protocol (APL 91 protocol), has proved extremely safe when applied on a multicenter basis (including numerous centers which diagnose 1 case of APL or less annually, as APL is such a rare disorder).

As seen above, dexamethasone has also proved useful in reversing the RA syndrome in the limited number of patients where it was used [13; Bollag, personal commun.], but probably has no effect on the leukocyte count itself. Its role will have to be precisely defined in larger cohorts of patients. Leukaphereses, as stated above, were not regularly useful in the New York experience and are sometimes difficult to carry out in emergency in some centers.

Partly in order to reduce the risk of hyperleukocytosis, we have recently treated 25 newly diagnosed and relapsing APL patients with lower doses of ATRA (25 mg/m^2/day). Results (21 CR in 25 patients) were identical to those obtained with conventional dose ATRA (45 mg/m^2/day), but the incidence of major hyperleukocytosis (8/25) was not reduced [Castaigne et al., unpublished].

As we reported above, mild therapeutic approaches such as low-dose Ara-C or hydroxyurea are insufficient in reversing rapidly rising leukocyte counts in APL in many patients. However, chemotherapy required to reduce the leukocyte count in such situations does not have to be as intensive as the 4+7 daunomycin-Ara-C course we used in our patients. In elderly patients particularly, where one wants to avoid a prolonged period of aplasia, 2 or 3 days of Ara-C at 100–200 mg/m^2/day, with or without 1 or 2 days of an anthracyclin, are generally sufficient. From our point of view, however, the purpose of administering high-dose chemotherapy (3+7 or 4+7 daunomycin-Ara-C combinations) as soon as hyperleukocytosis developed was not only to reduce leukocyte counts but also to provide further reduction in the leukemic burden in order to achieve cure.

Eradication of the Disease

Indeed, ATRA alone or followed by low-dose chemotherapy is incapable of curing APL. Our preliminary findings suggest that ATRA followed by intensive chemotherapy, by contrast, may cure a larger number of patients than conventional chemotherapy alone [14]. In this pilot phase of ATRA followed by intensive chemotherapy, patients received a 4+7 daunorubicin-Ara-C course after they had achieved CR with ATRA alone, or while still on ATRA if hyperleukocytosis rapidly developed. This was followed by reinduction courses and maintenance therapy. As seen above, 25 of the 26 (96%) treated patients achieved CR. In addition, the incidence of relapses was very low [14]. These results were compared with those of our previous APL 84 protocol with chemotherapy alone in newly diagnosed APL. In this protocol the CR rate was slightly lower (76%) and the incidence of relapses was higher, leading to a significant difference for actuarial disease-free interval and event-free survival between the APL 84 protocol and ATRA followed by chemotherapy [14]. These findings suggest that the latter may improve the outcome of newly diagnosed APL, by slightly increasing the CR rate, but perhaps more importantly by reducing the incidence of relapses. This hypothesis is now being tested in a randomized protocol comparing ATRA, followed by intensive chemotherapy and intensive chemotherapy alone in newly diagnosed APL. This protocol was started in April 1991, and 83 patients have so far been included. Preliminary results support an advantage for the ATRA followed by chemotherapy arm, although the difference is not significant for the moment. Early results of other groups [Warrell and Ohno, personal commun.] also suggest that ATRA followed by intensive chemotherapy is associated with very few relapses, although follow-up is also still limited in those experiences.

In relapsing APL patients, as seen above, subsequent administration of chemotherapy appears insufficient in eradicating the disease (with the possible exception of very intensive consolidation chemotherapy, which has not to our knowledge been administered to a substantial number of patients to draw any conclusion). Cure can so far only apparently be obtained after BMT, as in other types of acute myeloid leukemia relapsing after chemotherapy. In this situation particularly, overcoming progressive resistance to ATRA is a major goal.

Overcoming Resistance to ATRA

In order to prevent progressive resistance to ATRA, which leads to rapid relapse in most patients receiving ATRA alone, several approaches can be envisaged:

(1) Intermittent schedule of ATRA: in vitro and in vivo data suggest that relatively short exposure to ATRA may be sufficient in triggering the differentiation of most APL cells; by contrast, intermittent administration could avoid the increased catabolism observed with prolonged administration of ATRA, although this remains speculative.

(2) Synthesis of new retinoid compounds still capable of differentiating APL cells, but whose metabolism may not be modified during treatment, can be envisaged.

(3) Use of cytochrome P-450 antagonists. Because cytochrome P-450 seems to be involved in the increased catabolism of ATRA observed with prolonged administration, the use of cytochrome P-450 antagonists, such as ketoconazole, could be envisaged with the aim of overcoming resistance to ATRA.

(4) Combination of alpha-interferon to ATRA, in 1 patient who had become resistant to ATRA, restored in vivo efficacy of ATRA [31]. However, this effect of alpha-interferon was not confirmed by Warrel [personal commun.] in another patient.

Biological Tools for Monitoring Clinical Efficacy of ATRA

In vitro Study of Differentiation and Proliferation of APL Cells with ATRA

Huang et al. [11] and our group [10] have shown a close correlation between in vivo and in vitro differentiation with ATRA. Our preliminary results also suggest a correlation between in vitro proliferation (associated to their differentiation) of APL cells, and hyperleukocytosis in vivo [14]. Hyperleukocytosis can thus, to some extent, be anticipated by in vitro studies.

Pharmacokinetic Studies

Experience from our group suggests that patients not responding to ATRA have low serum peak levels of ATRA and high clearance rates [32]. As previously seen, Muindi et al. [24] found a progressive reduction of serum levels of ATRA with prolonged therapy, which could probably explain in part progressive resistance in patients. Thus, pharmacokinetics appear to be a useful parameter in the follow-up of patients who do not respond to ATRA and will be important in the assessment of strategies designed at overcoming resistance to ATRA.

Amplification of the Abnormal PML-RAR-Alpha Transcript by Reverse Polymerase Chain Reaction

At the RNA level, the fusion of PML and RAR-alpha genes resulting from t(15;17) translocation in APL leads to an abnormal transcript (of three possible types) [33–36]. This abnormal transcript can be amplified after reverse transcription to DNA, as for chronic myelocytic leukemia with the BCR-ABL fusion RNA. The high sensitivity of polymerase chain reaction (PCR) will then allow the detection of very small amounts of abnormal RNA and thus of minimal residual disease after treatment. Preliminary results with reverse PCR suggest eradication of the disease in allografted patients, but persistence of positive PCR after only one course of intensive chemotherapy or after ATRA alone [33–37]. The number of patients tested so far, however, is still very small. Reverse PCR will be important in determining for instance if better eradication of the disease is obtained after ATRA+chemotherapy, as compared to chemotherapy alone. Of note is that, in order to perform reverse PCR in remission, one should first obtain an RNA sample at diagnosis, because of the different possible breakpoints in the PML-RAR-alpha fusion RNA.

Conclusion

Recent clinical experience with ATRA suggests that this drug has been able to induce a very high CR rate (approximately 90%) in newly diagnosed and first relapsing APL. That ATRA is active through differentiation and not cytotoxicity has been clearly shown.

A major side effect of ATRA is the risk of major hyperleukocytosis, which requires rapid therapeutic intervention in order to avoid complications (RA syndrome, thromboembolic events).

Once CR has been achieved, some form of intensive therapy is required to achieve cure. Preliminary results suggest that ATRA followed by intensive chemotherapy in newly diagnosed APL may be superior to conventional therapy alone. This question is currently being addressed in a randomized fashion in the APL 91 protocol. In APL relapsing after chemotherapy, however, ATRA followed by chemotherapy does not seem to allow cure of the disease, unless an allogeneic transplant is performed. This suggests that ATRA should be included in the front line therapy of newly diagnosed APL.

References

1 Bernard J, Weil M, Boiron M, et al: Acute promyelocytic leukemia: Results of treatment with daunorubicin. Blood 1973;41:489–496.
2 Marty M, Ganem G, Fischer J, et al: Leucémie aiguë promyélocytaire: Etude rétrospective de 119 malades traités par daunorubicine. Nouv Rev Fr Hématol 1984;24:371–378.
3 Kartarjian H, Keating M, Walters R: Acute promyelocytic leukemia: MD Anderson Hospital Experience. Am J Med 1986;80:789–797.
4 Cunningham I, Gee T, Reich L, et al: Acute promyelocytic leukemia: Treatment results during a decade at Memorial Hospital. Blood 1989;73:1116–1122.
5 Sanz M, Jarque I, Martin G, et al: Acute promyelocytic leukemia: Therapy results and prognostic factors. Cancer 1988;61:7–13.
6 Cordonnier C, Vernant JP, Brun B, et al: Acute promyelocytic leukemia in 57 previously untreated patients. Cancer 1985;55:18–25.
7 Fenaux P, Pollet JP, Vandenbossche L, Dupriez B, Jouet JP, Bauters F: Treatment of acute promyelocytic leukemia: A report on 70 cases. Leuk Lymphoma 1991;4:249–256.
8 Fenaux P, Tertian G, Castaigne S, et al: A randomized trial of amsacrine and rubidazone in 39 patients with acute promyelocytic leukemia. J Clin Oncol 1991;9:1556–1561.
9 Tallman M, Kwaan H: Reassessing the hemostatic disorder associated with acute promyelocytic leukemia. Blood 1992;73:543–553.
10 Chomienne C, Ballerini P, Balitrand N, Daniel MT, Fenaux P, Castaigne S, Degos L: All-trans-retinoic acid in promyelocytic leukemias. II. In vitro studies of structure-function relationship. Blood 1990;76:1710.
11 Huang M, Yu-Chen Y, Shu-Rong C, et al: Use of all-trans-retinoic acid in the treatment of acute promyelocytic leukemia. Blood 1988;72:567–572.
12 Chen ZX, Xue YQ, Zhang R, et al: A clinical and experimental study on all-trans-retinoic acid treated acute promyelocytic leukemia patients. Blood 1991;78:1413–1419.
13 Frankel S, Weiss M, Warrell R: A 'retinoic acid syndrome' in acute promyelocytic leukemia: Reversal by corticosteroids (abstract). Blood 1991;70(suppl 1):380a.
14 Fenaux P, Castaigne S, Dombret H, et al: All transretinoic acid followed by intensive chemotherapy gives a high complete remission rate and may prolong remissions in newly diagnosed acute promyelocytic luekemia: A pilot study on 26 cases. Blood, in press.
15 Degos L, Chomienne C, Daniel MT, Berger R, Dombret H, Fenaux P, Castaigne S: Treatment of first relapse in acute promyelocytic leukaemia with all-trans-retinoic acid. Lancet 1990;ii:1440.
16 Warrell R: Clinical and pharmacological aspects of ATRA treatment in APL. 5th Int Symp on Differentiation Therapy, Sardinia, 1992.
17 Castaigne S, Chomienne C, Daniel MT, Berger R, Fenaux P, Degos L: All-trans-retinoic acid as a differentiating therapy for acute promyelocytic leukemias. I. Clinical results. Blood 1990;76:1704.
18 Warrel RP, Frankel SR, Wilson MD, Miller H, et al: Differentiation therapy of acute promyelocytic leukemia with tretinoin (all-trans-retinoic acid). N Engl J Med 1991;324:1385–1393.
19 Elliott S, Taylor K, White S, et al: Proof of differentiative mode of action of all-trans-retinoic acid in acute promyelocytic leukemia using X-linked clonal analysis. Blood 1991;79:1916–1919.
20 Ohashi H, Ichikawa A, Takagi N, et al: Remission induction of acute promyelocytic leukemia by all-trans-retinoic acid: Molecular evidence of restoration of normal hematopoiesis after differentiation and subsequent extinction of leukemic clone. Leukemia 1992;6:859–862.
21 Scrobohaci ML, Dombret H, Ghorra P, et al: Disseminated intravascular coagulation (DIC) and/or primary fibrinolysis in acute promyelocytic leukemia: Effect of all-trans-retinoic acid treatment. Blood 1991;78(suppl):48a.
22 Wang ZY, Sun GL, Lu JX, et al: Treatment of acute promyelocytic leukemia with all-trans-retinoic acid in China. Nouv Rev Fr Hématol 1990;32:34–36.
23 Muindi J, Frankel SR, Miller SR, Jr, et al: Continuous treatment with all-trans-retinoic acid causes a progressive reduction in plasma drug concentrations: Implications for relapse and retinoid 'resistance' in patients with acute promyelocytic leukemia. Blood 1992;79:299–303.
24 Muindi JR, Frankel SR, Huselton C, et al: Clinical pharmacology of oral all-trans-retinoic acid in patients with acute promyelocytic leukemia. Cancer Res 1992;52:2138–2142.
25 Cornic M, Delva L, Guidez F, et al: Induction of retinoic acid-binding protein in normal and malignant human myeloid cells by retinoic acid in acute promyelocytic leukemia patients. Cancer Res 1992;52:3329–3334.
26 Toh CH, Winfield DA: All-trans-retinoic acid and side-effects. Lancet 1992;i:1239–1240.
27 Akiyama H, Nakamura N, Nagasaka S, Sakamaki H, Onozawa Y: Hypercalcaemia due to all-trans-retinoic acid. Lancet 1992;i:308–309.
28 Runde V, Aul C, Heyll A, Schneider W: All-trans-retinoic acid: Not only a differentiating agent, but also an inducer of thromboembolic events in patients with M_3 leukemia. Blood 1992;79:534–535.
29 Koik T, Tatewaki W, Aoki A, et al: Brief report: Severe symptoms of hyperhistaminemia after the treatment of acute promyelocytic leukemia with tretinoin (all-trans-retinoic acid). N Engl J Med 1992;327:385–387.
30 Fenaux P, Castaigne S, Chomienne C, Dombret H, Degos L: All-trans-retinoic acid treatment for patients with acute promyelocytic leukemia. Leukemia 1992;6:64–67.
31 Koller E, Krieger O, Kasparu H, Lutz D: Restoration of all-trans-retinoic acid sensitivity by interferon in acute promyelocytic leukaemia. Lancet 1991;ii:1154.
32 Lefebvre Ph, Thomas G, Gourmel B, et al: Pharmacokinetics of oral all-trans-retinoic acid in patients with acute promyelocytic leukemia. Leukemia 1991;5:1–5.
33 De Thé H, Lavau C, Marchio A, et al: The PML-RAR alpha fusion mRNA generated by the t(15;17) translocation in acute promyelocytic leukemia encodes a functionally altered RAR. Cell 1991;66:675–684.
34 Castaigne S, Balitrand N, De Thé H, et al: A PML/retinoic acid receptor alpha fusion transcript is constantly detected by RNA-based polymerase chain reaction in acute promyelocytic leukemia. Blood 1992;79:3110–3115.
35 Miller WH Jr, Kakizuka A, Frankel SR, et al: Reverse transcription polymerase chain reaction for the rearranged retinoic acid receptor alpha clarifies diagnosis and detects minimal residual disease in acute promyelocytic leukemia. Proc Natl Acad Sci USA 1992;89:2694–2698.
36 Diverio D, Lo Coco F, D'Adamo F, et al: Identification of DNA rearrangements at the retinoic acid receptor-alpha (RAR-alpha) locus in all patients with acute promyelocytic leukemia (APL) and mapping of APL breakpoints within the RAR-alpha second intron. Blood 1992;79:3331–3336.

Acta Haematol 1993;89(suppl 1):28–32

B. Dreno
P. Celerier
P. Litoux

Department of Dermatology,
Hôtel Dieu, Nantes, France

Roferon-A in Combination with Tigason in Cutaneous T-Cell Lymphomas

Key Words
Cutaneous lymphomas
Interferon-α
Retinoids
Therapy

Abstract
In cutaneous T-cell lymphomas (CTCL; mycosis fungoides and Sézary syndrome), the standard therapies tend to be effective but not curative. Single drug therapy with either interferon-α or retinoids shows a response rate of about 45%. In this article, we report the results obtained in the treatment of CTCL with a combined therapy. To date, four clinical studies have been carried out using a combination of low doses of interferon-α and retinoids for treatment in the early stages of CTCL. The mechanism of action of this combination therapy is unknown.

Diseases

Cutaneous T-cell lymphomas (CTCL) represent a group of malignant lymphomas that clinically originate in the skin, but after a variable period of time, may progress to involve lymph nodes and visceral organs. The best-known representatives of this group of malignant lymphomas are mycosis fungoides (MF) and Sézary syndrome (SS). Both are characterized by a proliferation of T lymphocytes mainly of the helper (CD4) type, which infiltrate the epidermis and papillary dermis in the early stages of disease. The therapeutic regimens for MF are often very effective but generally noncurative and are mostly directed at the skin lesions. In the early stages of the disease even topical glucocorticosteroid treatment may be effective. Other treatments are topical application of nitrogen mustard in solution, psoralen combined with ultraviolet A irradiation, total skin X-ray and electron beam irradiation. Chemotherapy is only used in advanced stages of the diseases. The lack or curative therapy has initiated an ongoing search for new systemic therapeutic approaches.

Retinoids

Within the past 10 years, biological response modifiers (retinoids and interferon-α) have been introduced in the treatment of CTCL. In seven studies [1–7], 75 patients were treated with retinoids alone; response rates of 21% complete remission and 36% partial remission were obtained (table 1). Retinoids were given daily with a dosage of between 0.5 and 1.5 mg/kg. Results were similar with both etretinate and 13-*cis*-retinoic acid. Clinical response occurred in the plaque and tumor stages of the disease but was less evident with SS. In the majority of patients, treatment appeared to be merely palliative and not curative,

We thank Miss Arlette Bénardin for her excellent technical assistance.

Dr. Brigitte Dreno
Department of Dermatology
Hôtel Dieu
Pl. A.-Ricordeau
F–44035 Nantes Cedex 01 (France)

0001–5792/93/0895–0028
$ 2.75/0

since dermal infiltrate never completely cleared. Side effects consisted primarily of dryness of skin and mucous membranes, and were usually controlled with emollients or reduction of retinoid doses. Other problems involved increased triglyceride levels and liver enzymes, which were reversible after discontinuation of therapy.

Interferon-α

Interferon-α has recently been introduced in the treatment of CTCL. A review of 11 studies [8–18] involving 183 patients (table 2) showed 22% with complete remission and 41% with partial remission after treatment. The dosages varied from between 3 to 50×10^6 IU, depending on the study. Very high dosages, given in the earlier studies [8, 11], induced severe side effects and treatment often had to be discontinued. Currently, a dosage of between 3 to 9×10^6 IU is recommended. After 2 months of daily therapy, interferon-α is usually given 3 times weekly. In our experience, we have had no better results with the higher dosages than with a dosage of between 3 to 9×10^6 IU. Side effects (e.g. mild to moderate flu-like symptoms, leukopenia, weight loss, depression) are usually bearable.

It must be stressed that two kinds of clinical responders to interferon-α therapy may be distinguished. In some patients, the clinical effect of interferon occurs quickly (in less than 2 months), but in other patients, it may not be observed before 3 or 4 months of treatment. In our study, 6 patients went into complete remission only after 10 months of treatment. Interferon-α appears to induce a temporary remission in advanced stages of CTCL, but it is not a sustained, complete remission. The response is more favourable in stages Ib–IIb. At the moment, we have 7 patients in complete remission who have been without further therapy for between 2 and 4 years.

The results obtained by using either retinoids or interferon-α alone led to further investigation of the therapeutic effect and toxicity profile of interferon-α in a combination treatment with retinoids.

Combination Therapy

Four studies have been performed using this combination (table 3). In a pilot study of 7 patients, Thestrup-Pederson et al. [13] observed 1 complete remission, 3 partial remissions and 1 progressive disease. Two patients were nonevaluable. The evaluable patients were in stages II, IVa, and IVb. The treatment schedule initially consisted of 3×10^6 IU of interferon alfa-2a for 3 days, and then increased every 4 days to 9 and 18×10^6 IU, combined with etretinate 0.7 mg/kg. After 3 months, interferon alfa-2a was given 3 times weekly. Interestingly, the patient in complete remission remained so for 18 months without any treatment.

In the second study [19], 6 patients (stages II–IVb) were given a combined daily regimen of etretinate 50 mg and 3×10^6 IU interferon alfa-2a subcutaneously for 3 months, and then 3 times per week for 6 months, followed by an increase to 9 or 18×10^6 IU depending on response and tolerance. Three patients were in complete remission, 1 in partial remission, 1 in progressive disease, and treatment was discontinued in 1 patient due to the development of autoimmune hemolytic anemia. More recently with 7 patients (stages Ia–IVb), Knobler et al. [20] obtained 2 complete and 2 partial remissions using low dosages of interferon alfa-2b subcutaneously and 13-*cis*-retinoic acid (1 mg/kg) 3 times per week. After 6 months, dosages were reduced by 50% and final maintenance therapy consisted of interferon alfa-2b alone. These patients showed no relapse for up to 15 months. The average time to respond to therapy was 3 months.

The last report is a multicentric study [18] including 45 patients in stages Ia–IVb. For this study, patients first received 6×10^6 IU interferon alfa-2a alone, daily, for 2 months. After 2 months, patients in partial or complete remission received 6×10^6 IU interferon alfa-2a 3 times weekly for the next 10 months, and for nonresponders, interferon alfa-2a was increased to 9×10^6 IU/day for another month. After 3 months of interferon alfa-2a therapy, etretinate (0.5 mg/kg/day) was added in the treatment regimen of nonresponders.

After 3 months of interferon alfa-2a therapy alone, there were 19 (42.3%) responders (complete and partial remissions). Etretinate was given to 26 patients. After 12 months of treatment, 10 patients stopped treatment: 7 had progressive disease (3 stage IVa, 4 SS), 2 had side effects (thrombocytopenia, threefold increase in alkaline phosphatase levels), and 1 patient refused to continue therapy. Of the remaining 35 patients, 28 were responders (13 complete and 15 partial) and 7 were nonresponders. Of the 28 responders, 17 were finally treated with interferon alfa-2a alone (60.1%) and 11 (39.9%) had interferon alfa-2a and etretinate. Etretinate was added after 3 months of interferon alfa-2a therapy in 26 patients (54%) showing no response. After 9 months with this combination therapy, 11 patients (46%) showed a good response (fig. 1a, b). It is important to note that the results of this study were based on the clinical stage of the disease (table 4). Among the 25 patients with MF stages I and II, there were 20 responders (80%) after 12 months of treatment: 12 receiving interferon alfa-2a alone and 8 receiving interferon alfa-2a and etreti-

Table 1. Effects of retinoids therapy alone in CTCL

Studies	n	Stages	Doses mg/kg	Responses	
				PR	CR
Claudy et al. [2]	1	II	1	1	
Claudy et al. [3]	12	4 PP, 2 NEL	0.8	2	5
Souteyrand et al. [1]	5	3 PP, 2 II	1	1	2
Kessler et al. [7]	25	II–III	1–2	8	3
Wargon and Downie [5]	1	II	2	1	
Thomsen et al. [6]	24	I–IV	1–2	11	6
Warell et al. [4]	7	II	1	3	0
Total	75			27 (36%)	16 (21%)

PR = Partial remission; CR = complete remission; PP = parapsoriasis; NEL = nonepidermotropic lymphoma.

Table 2. Effects of therapy with interferon-α alone in CTCL

Studies	n	Stages	Doses	Responses	
				PR	CR
Bunn et al. [8]	19	II–IVb	50×10^6 IU	6	3
Lang et al. [9]	5	III–IV	$3–18 \times 10^6$ IU	1	0
Covelli et al. [10]	19	I–IVa	$3–18 \times 10^6$ IU	9	8
Ihde et al. [11]	21	II–IVb	$3–18 \times 10^6$ IU	6	0
Tura et al. [12]	15	II–IV	$3–18 \times 10^6$ IU	8	1
Thestrup-Pedersen et al. [13]	5	II–IVb	$3–36 \times 10^6$ IU	3	0
Hagberg et al. [14]	4	?	$2–4 \times 10^6$ IU	1	1
Olsen et al. [15]	22	Ia–IVa	$3–36 \times 10^6$ IU	10	3
Estrach et al. [16]	9	Ib–IVa	5×10^6 IU	4	2
Papa et al. [17]	45	I–IV	$3–18 \times 10^6$ IU	21	11
Dreno et al. [18]	19	Ia–IVb	$6–9 \times 10^6$ IU	7	12
Total	183			76 (41%)	41 (22%)

PR = Partial remission; CR = complete remission.

Table 3. Effects of combination therapy with interferon and retinoids in CTCL

Studies	n	Stages	Doses	Responses	
				PR	CR
Thestrup-Pedersen et al. [13]	7	II–IVb	$3–18 \times 10^6$ IU 0.7 mg/kg etretinate	3	1
Braathen and McFadden [19]	6	II–IV	$3–18 \times 10^6$ IU 50 mg/kg etretinate	1	3
Knobler et al. [20]	7	Ia–IVb	3×10^6 IU 1 mg/kg 13-*cis*-retinoic acid	2	2
Dreno et al. [18]	26	Ia–IVb	$6–9 \times 10^6$ IU 0.5 mg/kg etretinate	7	4

PR = Partial remission; CR = complete remission.

nate. Of the 20 patients with MF stage IVb or SS, only 8 showed a response (40%, $p>0.001$): 5 receiving interferon alfa-2a alone (2 MF, 3 SS) and 3 receiving interferon alfa-2a and etretinate (1 MF, 2 SS). The combination of interferon alfa-2a and etretinate was given to 12 patients in stages I and II with good results in 8 (67%) and to 14 patients in MF stage IVb or SS with only 3 good results (21%). Two patients with stable disease after 3 months of interferon alfa-2a therapy were not given etretinate because of cardiac problems and interferon alfa-2a (9×10^6 IU) was continued without notable change.

Thus, this multicentric study [18] confirms the results of three other reports [13, 19, 20] that interferon-α in combination with retinoid causes significant improvement in some patients with CTCL. This improvement is obtained with low-dose regimens of interferon-α therapy, which is an important point to consider because of the tolerance of the relatively mild side effects of this combined treatment: skin dryness, flu-like symptoms, arthralgia, increase in the levels of liver enzymes, triglycerides and cholesterol. In our study, low doses of interferon-α permit long-term use of this combined treatment with minor side effects. Furthermore, it has been our experience [confirmed by ref. 20] that after 6 months of continuous therapy, doses may be reduced by 50%. In our study, 5 patients (stages II, III) discontinued treatment after 12 months of the combined therapy, and showed no relapse for up to 12 months.

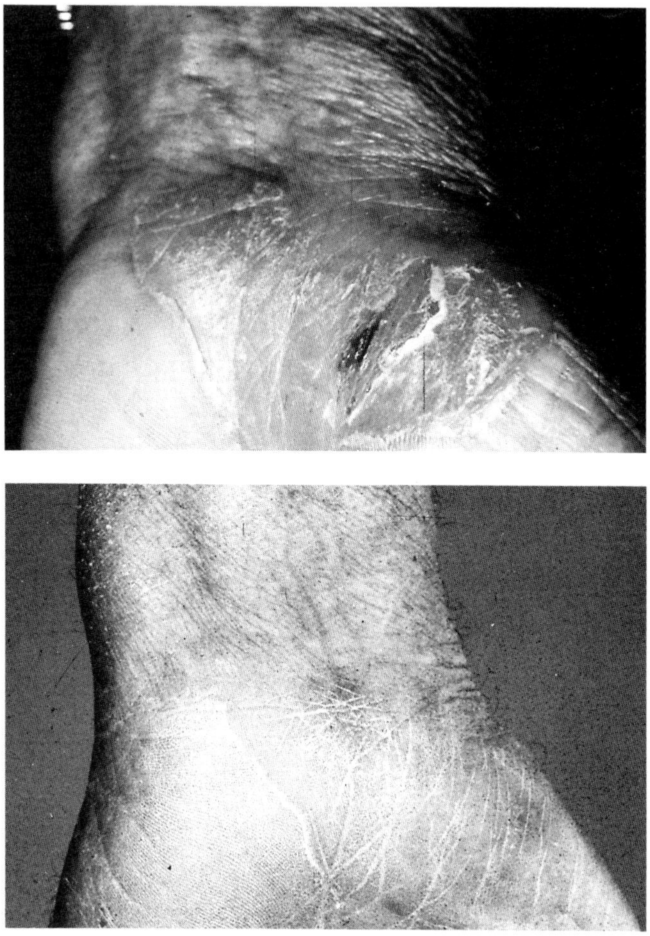

a

b

Fig. 1. MF stage II before treatment (**a**) and after (**b**) 2 months of combined therapy (Roferon A 6×10^6 IU and etretinate 1 mg/kg/day).

Mechanism of Action

The mechanism of action of this combined therapy remains unclear. Interestingly, it appears that retinoids and interferon-α have similar points of action: keratinocytes, lymphocytes and tumoral cells. Interferon-α and retinoids have an anti-proliferative effect on the keratinocytes, modulating their biological response [21, 22]. These two agents are able to increase cytotoxic activity of T cells and natural killer cells which might result in an enhanced cell-mediated lysis of malignant target cells [23, 24]. Finally, these biological response modifiers also have modulating enhancing effects on the expression of cell surface molecules such as histocompatibility and tumour-associated antigens [23, 25]. Moreover, a modulation of interferon-α antiviral activity by retinoic acid has been demonstrated in vitro [26].

In conclusion, because of the efficacy and the low incidence of severe side effects observed in these four studies, a combination of low-dose interferon-α and retinoids should be considered, in addition to the already well-established treatment schedules, in the therapy of CTCL.

Table 4. Results based on the clinical stages of disease with the combined therapy in the 45 patients of the multicentric study [21]

	Early stages I, II		Advanced stages III, IV
IFN alone, IFN + ET			
Good results	20		8
		$p<0.01$	
Poor results	5		12

	Early stages I, II		Advanced stages III, IV	
	good results	poor results	good results	poor results
IFN	12	1	5	1
IFN + ET	8	4	3	11

Good results = Complete or partial remission; poor results = stabilisation or progressive disease; IFN = interferon-α; ET = etretinate.

References

1 Souteyrand P, Thivolet J, Fulton R: Treatment of parapsoriasis en plaque et mycosis fungoides with an oral aromatic retinoid (RO 10-9359). Retinoids 1978;1:313–316.

2 Claudy AL, Delomier Y, Hermier C: Treatment of cutaneous T-cell lymphoma with a new aromatic retinoid (RO 10-9359). Arch Dermatol Res 1982;273:37–42.

3 Claudy AL, Rouchouse B, Boucheron J, Le Petit JC: Treatment of cutaneous lymphoma with etretinate. Br J Dermatol 1983;109: 49–56.

4 Warell RP Jr, Coonley CJ, Kempin JJ, Myskowski P, Safai B: Isotretinoin in cutaneous T-cell lymphoma. Lancet 1983;ii:629.

5 Wargon O, Downie D: A case of mycosis fungoides treated with etretinate. Aust J Derm 1984;25:77–79.

6 Thomsen K, Molin L, Volgen G, Lange Wantzin G, Hellve L: 13-cis-Retinoic acid effective in mycosis fungoides. Acta Derm Venereol 1984;64:563–566.

7 Kessler JP, Jones SE, Levine N, Lynch PJ, Booth AR, Meyskens FL: Isotretinoin and cutaneous helper T-cell lymphoma (mycosis fungoides). Arch Dermatol 1987;123:201–204.

8 Bunn PA, Ihde DC, Foon KA: The role of recombinant interferon alpha-2a in the therapy of cutaneous T-cell lymphomas. Cancer 1986; 57:1689–1695.

9 Lang MH, Altmeyer P, Lodemann E, Holzmann H: Alpha-interferon in the therapy of cutaneous T-cell lymphomas. Z Hautkr 1986; 61:599–608.

10 Covelli A, Cavaliere R, Coppola G: Recombinant leukocyte A interferon (IFN-rA) as initial therapy in mycosis fungoides and Sézary syndrome (abstract). Proc ASCO 1987;6:189.

11 Ihde DC, Stays R, Sausville EA: Phase II trial of intermittent high-dose recombinant interferon alpha-2a in mycosis fungoides and Sézary syndrome. Proc Am Assoc Cancer Res 1987, p 208.

12 Tura S, Mazza P, Zinzani PL, Ghetti PL, Poletti G, Gherlinzoni F, Montagnani A, Criscuolo D: Alpha recombinant interferon in the treatment of mycosis fungoides. Haematologica 1987;72:337–340.

13 Thestrup-Pedersen K, Hammer R, Kaltoft K, Sogard H, Zachariae H: Treatment of mycosis fungoides with recombinant interferon-alpha-2a alone and in combination with etretinate. Br J Dermatol 1988;118:811–818.

14 Hagberg H, Juhlin L, Scheynius A, Tjernlund U: Low dosage alpha-interferon treatment in patients with advanced cutaneous T-cell lymphoma. Eur J Haematol 1988;40:31–34.

15 Olsen EA, Rosen ST, Vollmer RT, Variakojis D, Roenigk HH Jr, Diab N, Zeffren J: Interferon alfa-2a in the treatment of cutaneous T-cell lymphoma. J Am Acad Dermatol 1989; 20:395–407.

16 Estrach T, Vicente MA, Marti R, Lecha M, Mascaro JM: Tratamiento de linfomas cutanos de células T con dosis bajas de interferon alfa 2b recombinante: Estudio de nueve casos. Med Cut IL.A 1990;18:354–358.

17 Papa G, Tura S, Mandelli F, Vegna ML, Defazio D, Mazza P, Zinzani PL, Simoni R, DePita O, Ferranti G, Ghetti PL, Criscuolo D: Is interferon alpha in cutaneous T-cell lymphoma a treatment of choice? Br J Haematol 1991;79 (suppl):48–51.

18 Dreno B, Claudy A, Meynadier J, Verret JL, Souteyrand P, Ortonne JP, Kalis B, Godefroy WY, Beerblock K, Thill L: The treatment of 45 patients with cutaneous T-cell lymphoma with low doses of interferon-α2a and etretinate. Br J Dermatol 1991;125:456–459.

19 Braathen LS, McFadden N: Successful treatment of mycosis fungoides with the combination of etretinate and human recombinant interferon alfa-2a. J Dermatol Treatment 1989;1:29–32.

20 Knobler RM, Trautinger F, Radaszkiewicz T, Kokoschka EM, Micksche M: Treatment of cutaneous T-cell lymphoma with a combination of low-dose interferon alfa-2b and retinoids. J Am Acad Dermatol 1991;24:247–252.

21 Orfanos CE: Oral retinoids. Br J Dermatol 1980;103:473–481.

22 Yaar M, Karassid RL, Schnipper LE, Gilchrest BA: Effects of alpha and beta interferons on cultured human keratinocytes. J Invest Dermatol 1985;85:70–74.

Acta Haematol 1993;89(suppl 1):33

John M. Goldman

Closing Remarks

The speakers have very eloquently conveyed the message of this satellite symposium. We have in the last decade seen the birth and development of a completely new approach to treating patients with hematological malignancies – an approach that exploits what we know of the basic lesion in leukemia and lymphoma and tries to correct the cellular or molecular defect by recruiting normal or near-normal mechanisms. The era of cytotoxic drugs is not finished, rather we may hope that the combined use of cytotoxic drugs and biological molecules, some of which we have heard about and others of which are yet undiscovered, will increase the chance of complete remission and cure for our patients. I see no reason why this should not be.

Author Index

Subject Index

New books of the American Association of Blood Banks

New books

Blood Groups: Chido/Rodgers, Knops/McCoy/York and Cromer

Editors: Moulds, J.M. (Houston, Tex.); Laird-Fryer, B. (Dallas, Tex.)
X + 66 p., 9 fig., 8 tab., hard cover, 1992
SFr. 118.– / DM 141.– / £ 59.00
Prices subject to change
DM price for FRG, £ price for UK only
ISBN 3–8055–5728–0
Distributed in USA and Canada exclusively by the AABB

In the past they were referred to as 'high-titer, low-avidity (HTLA)' antibodies. Now, many of the antigens recognized by these HTLA antibodies have been located on complement-related proteins. This new publication reviews the historical aspects of how these antibodies were found and named; their protein structures, their function and genetic makeup of the corresponding genes. Disease associations including PNH, SLE and other autoimmune diseases are also covered. Gain a greater insight into the serologic workup and importance of these antibodies in transfusion medicine!

Clinical Decisions in Platelet Therapy

Editors: Kurtz, S.R. (Burlington, Mass.); Brubaker, D.B. (Spokane, Wash.)
XII + 128 p., 4 fig., 13 tab., hard cover, 1992
SFr. 124.– / DM 148.– / £ 62.00
Prices subject to change
DM price for FRG, £ price for UK only
ISBN 3–8055–5729–9
Distributed in USA and Canada exclusively by the AABB

This publication addresses one of the fastest growing aspects of transfusion medicine today. Subjects to be discussed include: Determining the Dose of Platelets for Transfusion; Single Donor versus Random Donor Platelet Support; Mechanisms of Allo-(R)immunization to HLA Class I Antigens. Also covered: gaining a better understanding of alloimmunization and the refractory state; the basic immunology associated with alloantibody response in non-immunized patients and postulated mechanisms (including supportive data) in HLA alloimmunization that result in patients becoming refractory to platelet transfusion; the clinical and laboratory approach to the diagnosis of the refractory state; and transfusion practices and the clinical management of patients who are refractory to platelet transfusions and methods to prevent alloimmunization and the refractory state.

Fields of Interest
Hematology; Biochemistry, Genetics, Immunology, Laboratory, Serology, Transfusion

Available from Karger
Please ask for detailed prospectus

KI 93547

KARGER